TABLE OF CONTENTS

Acknowledgments...III
Introduction..IV

Chapter 1
WORKING WITH RATIONAL NUMBERS

Lesson	Page
1. Number Sense	2
2. Operation on Integers	19
3. Order of Operations	53

Chapter 3
EQUATIONS AND INEQUALITIES

Lesson	Page
1. Equations	170
2. Inequalities	184
3. Proportions	205

Chapter 2
EXPRESSIONS

Lesson	Page
1. Vocabulary Development	79
2. Evaluating Expressions	96
3. Simplifying Expressions	106
4. Exponents	138

Chapter 4
LINEAR FUNCTIONS AND VERTICAL LINES

Lesson	Page
1. Coordinate Plane Vocabulary	222
2. Graphing Lines	241
3. Writing Equations of Lines	273

TABLE OF CONTENTS CONT...

Chapter 5
LINEAR SYSTEMS

Lesson	Page
1. Linear Systems	286
2. Systems of Linear Inequalities	306

Chapter 6
POLYNOMIALS

Lesson	Page
1. Developing Vocabulary	322
2. Operations on Polynomials	331
3. Factoring and Solving Polynomial Equations	346

Chapter 7
RADICALS

Lesson	Page
1. Simplifying Radicals	368
2. Operations with Radicals	374
3. Applications	381

Chapter 8
QUADRATIC FUNCTIONS

Lesson	Page
1. Vocabulary Development	388
2. Graphing Quadratic Functions	406
3. Solving Quadratic Functions	435

Cooperative Learning and Algebra 1: Becky Bride
Kagan Publishing • 1 (800) 933-2667 • www.KaganOnline.com

ACKNOWLEDGMENTS

Many people have helped to make this book possible. Thank you to my family, who understood when I took time away from them to write and edit. My children were kind when I needed the computer and they had to stop socializing with their friends via AIM. My husband was incredibly patient when the editing took me away from the kitchen remodel that we were so engrossed in. Without my family's support and encouragement, this book would not be a reality. Thank you to Michelle Kuebel who was my student assistant in 2003-04 and painstakingly worked all the problems so the activities would have answers. Thanks to my 2003-04 Algebra 1 class who did all of the investigations and told me how to word them so that a student would better understand the directions. Most of all, thank you to my Lord who gave me the gifts and talents to teach and write and gave me the imagination to see how algebra concepts could be made concrete.

To Spencer and Laurie Kagan, thank you for introducing me to the structural approach to cooperative learning, which is the foundation of this book. It has given me an opportunity to share with other teachers how the structural approach to cooperative learning can be successfully implemented in the mathematics classroom. Most of all, thank you for the tools to teach effectively. Because of these tools, I love teaching more today than when I first began over 18 years ago. Rather than experiencing burnout, I have been invigorated in my teaching and for that I am truly thankful.

Many thanks to the staff at Kagan Publishing, who took a rough draft and made it look beautiful. The creative book layout was designed by Denise Alphabet. Kim Fields edited the text. The icons, layout, and technical illustrations were the brainstorm of Andrew Bartel. The cover was illustrated by Erin Kant. Thank you to Jennifer McDonald for reviewing the book. Her suggestions for changes, corrections, and comments were very helpful during the editing phase. A special thank you to Miguel Kagan and Becky Herrington, who were my consultants, mentors, and project managers. I appreciated your encouragement and support.

A very special thank you to all the teachers who will use this book so their students' experience with Algebra 1 will be rich, engaging, and fun-filled.

INTRODUCTION

Have you ever been frustrated with your students because they don't seem to understand what you have taught and retaught several times? Do you ever feel like you are teaching rocks rather than young people? Do you go home at the end of a day and ask, "Why did I even come?" I have felt that way many times and went home asking myself those questions. The number of children that would come after school for help because they did not understand what had been taught ranged from 10 to 20 students a day. I knew there had to be a better way—I just did not know how. I knew that I could not change my students, nor could I pick and choose who I taught. That was when I decided that I was the one who would have to be transformed. In my quest for more effective ways of teaching, I discovered Kagan Cooperative Learning. I had put students together before to work on a project and it was disaster. I did not know why, and swore that I would never do that again. What amazed me about the structural approach to cooperative learning was how it ensured equal participation, required dependence of the group members on each other, and held the students individually accountable. As I investigated further, I was sold.

The human mind can only hold so much information before it goes into overload. In overload mode, all that it hears never gets past the ears because the brain is full. So in a 30-minute lesson, my students were with me for the first 10 minutes, then they went into overload. No wonder they did not know how to do the homework. With the structural approach, a lesson is broken down into 10-minute increments. Students process the information they received with an activity that uses a Kagan structure at the end of each increment. Because the brain has processed this information, it is ready to receive new information so the next part of the lesson can be absorbed. Because the structures ensure positive interdependence among the students, equal participation, individual accountability, and at least 25 percent of the class actively processing at once, there are no "hogs" (students who want to do it all) and no "logs" (students who won't do anything). Now the only students who stay after school are the ones who were absent and missed the lesson. The transformation has been amazing, and my passion, enthusiasm, and zeal for teaching has returned. After my first 5 years of teaching, I was burned out. I left teaching for 8 years and 2 years after I returned, I found the structural approach. That was 10 years ago. Instead of burnout I actually love going to school each day. My students enjoy it as much as I do because it adds variety in the classroom, they bond with their teammates, and they are more successful than they have ever been in mathematics. It is a win-win situation for all.

Besides the activities to process the Algebra 1 curriculum, exploratory exercises are included for the students to discover most of the algebra concepts. I taught geometry for years and was able to develop hands-on investigations using compasses, straightedges, and patty paper. The curriculum could be made so concrete for the students that it was much easier for them to understand. Because I had taught geometry for so long, I wanted a change and chose to teach Algebra 2. Be careful what you ask for! After a month in Algebra 2, I realized that algebra was very abstract. I missed the concrete investigations and went back to geometry the next year. Several years later, I was given the opportunity to develop a non-advanced placement calculus class. The book I chose to teach the class gave me the insight I needed to make discovery activities for algebra. This text presented

Cooperative Learning and Algebra 1: Becky Bride
Kagan Publishing • 1 (800) 933-2667 • www.KaganOnline.com

everything from a numeric, graphic, and algebraic perspective. I know that we teach all of these in Algebra 1, but I believe that we have failed to show the students how intricately related they are. The fact that I didn't even see this until 15 years into my teaching and after majoring in mathematics education, shows that we have failed to link these perspectives sufficiently. This revelation began the transformation of the algebra curriculum. Exploratory activities could be developed for students to discover the order of operations, rules for exponents, definitions of zero and negative exponents, linear function concepts, and even more. Some of the investigations have students explore numeric patterns or graphical patterns and then develop a conjecture based on the patterns (inductive reasoning at its best). Some begin with an exploration of the graphical perspective that leads to an understanding of the numeric perspective, which allows for the generalization to the algebraic perspective. The teacher is transformed from "giver of all information" to facilitator of what students have discovered. When students discover concepts—the concepts are theirs. It is well worth the time. Because these explorations use inductive reasoning, the student may feel that by the end of the exploration, it is getting repetitive. Part of this is inherent in inductive reasoning. I have tried to give enough problems so that the struggling students will be able to see the pattern.

Another revelation that I had was vocabulary development. It is a no-brainer that vocabulary development in a geometry class is essential because so much of the geometry is rooted in definitions. Vocabulary can't be ignored. It did not occur to me that the vocabulary development in the algebra classes is as critical as it is in the geometry classes. Sure I defined "exponent," "coefficient" and such. I assumed that the students were fluent with the vocabulary without putting activities in place to ensure its development. It wasn't until I was doing my portfolio for National Board for Professional Teaching Standards that I realized that my Algebra 2 students did not know their vocabulary! They were doing an activity that required them to write about their understanding. They used "x-axis" when they referred to an "x-coordinate" and "y-axis" for "y-coordinate." I was shocked since this was Algebra 1 vocabulary. Vocabulary development doesn't happen unless we ensure that it happens. With state and national tests requiring students to be able to communicate mathematically, we must take seriously vocabulary development. Included in this book are exploratory activities for the students to look at examples and non-examples of a vocabulary word and develop their own definition, which will ultimately result in a definition refined by the class. After these explorations follow cooperative learning activities designed for students to process the definitions. Because all of the exploratory activities require students to write about their mathematical discoveries, this vocabulary development is essential.

The book is divided into 8 chapters that cover the core concepts of the Algebra 1 curriculum. Many of the concepts are also taught in pre-algebra classes and could be used for those also. Each chapter is broken down into lessons, and each lesson contains exploratory and cooperative learning activities. Each processing activity uses a Kagan Cooperative Learning structure. At the beginning of each chapter is a detailed list of each lesson and the activities it contains, along with a synopsis of the chapter. Each lesson also begins with a synopsis. Most of the lessons end with students generating a graphic organizer of what they have

INTRODUCTION CONTINUED...

learned, giving them an opportunity to synthesize their understanding. Following the synopsis of each lesson are teacher notes for the activities, the name of the structure used, materials required for the activity, and step-by-step directions on how to do the activity. The chapters, lessons, and activities are numbered sequentially. In the right-hand corner of each activity is a number. The first number designates the chapter, the second number designates the lesson, and the third number designates the transparency or blackline. This system makes navigating through the book a breeze. Answers are included on the transparencies, blacklines, or in the teacher's notes.

This book was written for the teacher. My hope is that this book opens a whole new way to look at the Algebra 1 curriculum—that through the vocabulary development, exploratory investigations, and cooperative learning activities to process the curriculum, your students will have a grasp of Algebra 1 in a way that is unparalleled. Enjoy!

Cooperative Learning and Algebra 1: Becky Bride
Kagan Publishing • 1 (800) 933-2667 • www.KaganOnline.com

WORKING WITH RATIONAL NUMBERS

This chapter begins with sets of numbers by providing a review of the sets the students have already worked with and introducing two new sets: integers and rational numbers. The concepts of absolute value, opposite of a number, and multiplicative inverse are introduced so they can be reinforced throughout the book. Through investigations, the rules for operations on integers are student-generated and then processed. The chapter ends with students generating the order of operation rules and processing these rules. Each lesson ends with each team producing a graphic organizer to summarize, analyze, and synthesize their understanding.

LESSON 1 NUMBER SENSE

ACTIVITY 1: Can You Define Me?
ACTIVITY 2: Definitions
ACTIVITY 3: Name My Sets
ACTIVITY 4: Name a Number
ACTIVITY 5: What Is My Absolute Value?
ACTIVITY 6: What Is My Opposite?
ACTIVITY 7: What Is My Multiplicative Inverse?
ACTIVITY 8: Give Me an Example
ACTIVITY 9: What Did We Learn?

LESSON 2 OPERATIONS ON INTEGERS

ACTIVITY 1: Exploring Addition of Integers
ACTIVITY 2: Addition Using Tiles
ACTIVITY 3: Addition Using Rules
ACTIVITY 4: Addition Practice
ACTIVITY 5: Adding with Multiple Addends
ACTIVITY 6: Exploring Subtraction of Integers
ACTIVITY 7: Subtraction Using Tiles
ACTIVITY 8: Subtraction Using Rules
ACTIVITY 9: Subtraction Practice
ACTIVITY 10: Can You Handle the Mixture?
ACTIVITY 11: Exploring Multiplication of Integers

ACTIVITY 12: Multiplication Using Tiles
ACTIVITY 13: Multiplication Using Rules
ACTIVITY 14: Processing Multiplication and Division
ACTIVITY 15: Operations on Integers
ACTIVITY 16: Operations on Rational Numbers
ACTIVITY 17: Operations on Integers in the Real World
ACTIVITY 18: What Did We Learn?

LESSON 3 ORDER OF OPERATIONS

ACTIVITY 1: Investigating Addition/Subtraction vs. Multiplication/Division
ACTIVITY 2: Processing Addition/Subtraction vs. Multiplication/Division
ACTIVITY 3: Write a Problem
ACTIVITY 4: Investigating Multiplication/Division vs. Exponents
ACTIVITY 5: Processing Multiplication/Division vs. Exponents
ACTIVITY 6: Write a Problem
ACTIVITY 7: Investigating the Role of Grouping Symbols
ACTIVITY 8: Processing Order of Operations
ACTIVITY 9: What Did We Learn?

LESSON 1
NUMBER SENSE

This lesson begins with a fun activity to develop the students' ability to compare and contrast two sets to find their similarities and differences in order to generate a definition. The skills the students use in this activity are essential for vocabulary development presented in this book. The next activity requires the students to apply the skills developed from the first activity to generate definitions involving sets of numbers.

Student-generated definitions take more time to develop than teacher-generated ones. Because the students' understanding of these definitions far exceeds teacher-generated ones, the time is well spent. Since the students will be exploring operations on integers, this is a perfect opportunity to review the students' understanding of the number sets they have worked with for years and learn two new number sets that are an integral part of Algebra 1. The definitions for "exponent" and "square root" are included here because the students need to understand these concepts when they explore order of operations. I chose to include square roots in this chapter so that the order of operations will not have to be amended once they are used frequently in the quadratic unit. "Absolute value" of a number, the "opposite" of a number, and the "multiplicative inverse" of a number are also defined in this chapter so students can work with these concepts

numerically before they have to work with them algebraically. The numerical exploration for the definitions and the processing of the opposite of a number will lay the foundation upon which subtraction of integers, the equation unit, and the polynomial unit can build. The numerical exploration and processing of the multiplicative inverse of a number will lay the foundation for solving equations involving coefficients of variables that are not 1.

To deepen the students' understanding of the sets of numbers and to push the students into higher-level thinking, the activity "Name a Number" requires the students to name a number that fits a set criteria. This will also help the students see the intersection of the sets of numbers they defined.

Linking the real world to algebra is incredibly important. When students see how algebra can be applied in the real world, they are more eager to learn. The activity "Give Me an Example" gives the students the opportunity to name situations where different numbers are used.

Finally, each lesson will end with each team producing a graphic organizer of the curriculum learned in the lesson. Again this is an attempt to get the students to think on a higher level and synthesize the material learned.

Cooperative Learning and Algebra 1: Becky Bride
Kagan Publishing • 1 (800) 933-2667 • www.KaganOnline.com

1 CAN YOU DEFINE ME?

Exploratory

Discussing the three parts necessary for a good definition will assist the students with creating quality definitions, ultimately saving time. The three parts are as follows:

1. Stating the term.

2. Stating the nearest classification.

3. Stating those items that make it unique.

As an example, examine the terms "multiplicative inverse" of a number and "additive inverse" of a number.

Multiplicative inverse of a number is

Additive inverse of a number is

At this point, the first part of defining a word is satisfied. The terms have been stated. When the second part is written, the definitions look like this:

Multiplicative inverse of a number is a number

Additive inverse of a number is a number

The definitions are exactly the same after part two is complete, and students can see the need for the third part. With the completion of part three, the definitions look like this:

Multiplicative inverse of a number is a number that when multiplied by the given number gives an answer that is equal to 1.

Additive inverse of a number is a number that when added to the given number gives an answer equal to 0.

The third part completes the definition.

Solo

1. Individually, each student defines each vocabulary word by comparing and contrasting the examples and counterexamples. Students could do this part for homework.

Team with RoundRobin Sharing and RoundTable Recording.

2. In turn, each student reads to his/her team the definition he/she wrote.

3. The team discusses how to define the vocabulary word, based on the definitions just shared. The team must come to consensus on how to define the term.

4. Once the team reaches consensus, student 1 records the team definition on the team paper.

> **Structure**
> • Solo Team Class

> **Materials**
> • 1 Blackline 1.1.1 and pencil per student
> • 1 sheet of paper and pencil per team

5. Repeat steps 2-4 for each vocabulary word, rotating the recording.

Class

6. Choose a team and a student at that team to read their team's definition of the first vocabulary word. (Team and student spinners work well here.)

7. Write the definition on the board. Ask the class if they agree with part one, then part two, and finally part three of the definition, reworking what they want to change.

8. Once everyone has agreed, including the teacher, then it is recorded as the definition the class will use. You may find that lower-level classes prefer a somewhat longer definition if it has more meaning for them. Honors classes will try to be as concise as possible.

9. Repeat steps 6-8 for the remaining words.

ACTIVITY

2 DEFINITIONS

▶ **Structure**
· Solo Team Class

▶ **Materials**
· 1 Blackline 1.1.2 and pencil per student
· 1 sheet of paper and pencil per team

Exploratory

Solo

1. Individually, each student defines each vocabulary word by comparing and contrasting the examples and counterexamples. Students could do this part for homework. Remind students to simplify all fractions and square roots before comparing and contrasting.

Team with RoundRobin Sharing and RoundTable Recording.

2. In turn, each student reads to his/her team the definition he/she wrote.

3. The team discusses how to define the vocabulary word, based on the definitions just shared. The team must come to consensus on how to define the term.

4. Once the team reaches consensus, student 1 records the team definition on the team paper.

5. Repeat steps 2-4 for each vocabulary word, rotating the recording.

Class

6. Choose a team and a student at that team to read their team's definition of the first vocabulary word. (Team and student spinners work well here.)

7. Write the definition on the board. Ask the class if they agree with part one, then part two, and finally part three of the definition, reworking what they want to change.

8. Once everyone has agreed, including the teacher, then it is recorded as the definition the class will use.

9. Repeat steps 6-8 for the remaining words.

ACTIVITY

3 NAME MY SETS

▶ **Structure**
· RallyRobin

▶ **Materials**
· Transparency 1.1.3

Numeric

1. Teacher poses multiple problems using Transparency 1.1.3.

2. In pairs, students take turns orally stating the sets of numbers to which their number belongs.

ACTIVITY

4 NAME A NUMBER

▶ **Structure**
· RallyRobin

▶ **Materials**
· Transparency 1.1.4

Numeric Higher-Level Thinking

1. Teacher poses multiple problems using Transparency 1.1.4.

2. In pairs, students take turns orally stating a number that fits the criteria.

Cooperative Learning and Algebra 1: Becky Bride
Kagan Publishing · 1 (800) 933-2667 · www.KaganOnline.com

ACTIVITY

5 WHAT IS MY ABSOLUTE VALUE?

Numeric

Setup:
Teacher copies Blacklines 1.1.5 and/or 1.1.6 and cuts the cards apart.

1. Stand Up-Hand Up-Pair Up

2. Partner A quizzes his/her partner asking what the absolute value of the number on his/her card is and why the answer he/she gave is the absolute value.

3. Partner B answers.

4. Partner A praises or coaches.

5. Switch roles.

6. Partners trade cards.

7. Repeat steps 1-6.

Management tip: To start and stop the mixing, music works well. When the music starts, students mix. When the music stops, the students pair.

▶ **Structure**
• Quiz-Quiz-Trade

▶ **Materials**
• Set of integer cards or rational number cards or a mixture of the two (Blacklines 1.1.5 and 1.1.6)

ACTIVITY

6 WHAT IS MY OPPOSITE?

Numeric

Setup:
Teacher copies Blacklines 1.1.5 and/or 1.1.6 and cuts the cards apart.

1. Stand Up, Hand Up, Pair Up

2. Partner A quizzes his/her partner asking what the opposite of the number on his/her card is.

3. Partner B answers.

4. Partner A praises or coaches.

5. Switch roles.

6. Partners trade cards.

7. Repeat steps 1-6.

Management tip: To start and stop the mixing, music works well. When the music starts, students mix. When the music stops, the students pair.

▶ **Structure**
• Quiz-Quiz-Trade

▶ **Materials**
• Set of integer cards or rational number cards (Blacklines 1.1.5 and 1.1.6)

Chapter 1: Working with Rational Numbers

Lesson One

ACTIVITY

7 WHAT IS MY MULTIPLICATIVE INVERSE?

▶ **Structure**
· Quiz-Quiz-Trade

▶ **Materials**
· Set of integer cards or rational number cards (Blacklines 1.1.5 and 1.1.6)

Numeric

Setup:
Teacher copies Blacklines 1.1.5 and/or 1.1.6 and cuts the cards apart.

1. Stand Up, Hand Up, Pair Up

2. Partner A quizzes his/her partner asking what the multiplicative inverse is of the number on his/her card.

3. Partner B answers.

4. Partner A praises or coaches.

5. Switch roles.

6. Partners trade cards.

7. Repeat steps 1-6.

Management tip: To start and stop the mixing, music works real well. When the music starts, students mix. When the music stops, the students pair.

ACTIVITY

8 GIVE ME AN EXAMPLE

▶ **Structure**
· Placemat Consensus

▶ **Materials**
· 1 large sheet of paper (placemat) per team
· 1 pencil per student

Connecting

1. Each team draws a rectangle in the center of a large sheet of paper, which will be the central team space and sections off the area around the large rectangle for individual team member space similar to the diagram below. In the central team space, each team separates it into 3 sections, one for each problem.

2. Teacher provides the team with a type of number (e.g., fractions from the set of rational numbers, whole numbers, integers).

3. All teammates respond simultaneously in their individual space, writing as many real-world applications of that type of number as they can in the time allotted.

4. Teammate 1 announces one item he/she has written.

5. Teammates discuss the item.

6. If there is consensus that the item is an application, Teammate 1 records his/her application in the team space of the placemat set aside for problem 1, seeking help with wording if necessary.

7. The process is repeated for one or more rounds using RoundRobin, so each teammate in turn suggests an application and records the team consensus.

8. Repeat steps 2-7 for another set of numbers.

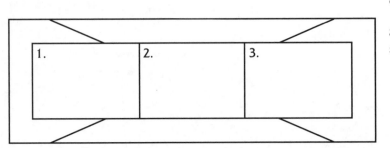

Cooperative Learning and Algebra 1: Becky Bride
Kagan Publishing · 1 (800) 933-2667 · www.KaganOnline.com

9 WHAT DID WE LEARN?

Synthesis via Graphic Organizer

1. Each teammate signs his/her name in the upper right corner of the team paper with the color pen/pencil he/she is using.

2. One teammate writes "Number Sense" in the center of the team paper in a rectangle.

3. Teammate 1 shares with the team one core concept he/she learned in the unit.

4. The student checks for consensus.

5. The teammates show agreement or lack of agreement with thumbs up or down.

6. If there is agreement, the students celebrate and the teammate records the core concept on the graphic organizer, connecting it with a line to the main idea, "number sense." If not, teammates discuss the response until there is agreement and then they celebrate.

7. Play continues with the next student's core concept, until all core concepts are exhausted.

8. Repeat steps 3-7 with teammates adding details to each core concept and making bridges between related ideas.

▶ **Structure**
• RoundTable Consensus

▶ **Materials**
• 1 large sheet of paper per team
• different color pen or pencil for each student on the team

Cooperative Learning and Algebra 1: Becky Bride
Kagan Publishing • 1 (800) 933-2667 • www.KaganOnline.com

Zingers

Examples	Counterexamples

Thingmabobs

Examples	Counterexamples

Cooperative Learning and Algebra 1: Becky Bride
Kagan Publishing • 1 (800) 933-2667 • www.KaganOnline.com

ACTIVITY
2 DEFINITIONS

For each word, compare the examples with the counter-examples and write a definition.

1. Square root of a number

Examples	Counterexamples

$\sqrt{4} = 2$; 2 was multiplied by itself two times

$\sqrt[3]{8} = 2$; 2 was multiplied by itself three times

$\sqrt{9} = 3$; 3 was multiplied by itself two times

$\sqrt[3]{27} = 3$; 3 was multiplied by itself 3 three times

$\sqrt{16} = 4$; 4 was multiplied by itself two times

$\sqrt[4]{16} = 2$; 2 was multiplied by itself 4 four times

$\sqrt{25} = 5$; 5 was multiplied by itself two times

$\sqrt[5]{243} = 3$; 3 was multiplied by itself five times

2. Natural numbers

Examples	Counterexamples

$1 \quad 10 \quad 531 \quad \dfrac{6}{3} \quad \dfrac{14}{7}$

$290 \quad 42 \quad 83 \quad 5{,}332$

$\sqrt{16} \quad \sqrt{4} \quad \sqrt{9}$

$-5 \quad 0 \quad 0.34 \quad -0.4.....$

$-7\dfrac{1}{2} \quad \dfrac{-4}{7} \quad 1\dfrac{5}{9} \quad \dfrac{2}{3}$

$\sqrt{5}$

3. Whole numbers

Examples	Counterexamples

$0 \quad 1 \quad 10 \quad 531 \quad \dfrac{6}{3} \quad \dfrac{14}{7}$

$290 \quad 42 \quad 83 \quad 5{,}332$

$\sqrt{16} \quad \sqrt{4} \quad \sqrt{9}$

$-5 \quad 0.34 \quad -0.4.....$

$-7\dfrac{1}{2} \quad \dfrac{-4}{7} \quad 1\dfrac{5}{9} \quad \dfrac{2}{3}$

$\sqrt{17} \quad \sqrt{14} \quad \sqrt{6}$

$-0.555555....$

ACTIVITY

2 DEFINITIONS

4. Integers

Examples	Counterexamples

Examples

−45   36   0   −10

290   −432   83   −2,331

$\sqrt{16}$ $-\sqrt{4}$ $\sqrt{9}$

Counterexamples

0.34   −0,4..... π

$-7\frac{1}{2}$ $\frac{-4}{7}$ $1\frac{5}{9}$ $\frac{2}{3}$

$\sqrt{17}$ $\sqrt{14}$ $\sqrt{6}$ $\sqrt{5}$

−0.44444....

5. Rational numbers

Examples

−32   0   4.5   −0.578

$-7\frac{1}{2}$ $\frac{-4}{7}$ $1\frac{5}{9}$ $\frac{2}{3}$ 103

$\sqrt{16}$ $\sqrt{4}$ $-\sqrt{9}$ 0.323232323....

Counterexamples

$-\sqrt{17}$ $\sqrt{14}$ $\sqrt{6}$

0.12112111211112... π

−5.223222332222333...

6. Opposite of a number

Examples

5 is the opposite of −5.

−3 is the opposite of 3.

−10 is the opposite of 10.

8 is the opposite of −8.

Counterexamples

−8 is not the opposite of −8.

4 is not the opposite of 4.

6 is not the opposite of 8.

−0.5 is not the opposite of −0.5.

Cooperative Learning and Algebra 1: Becky Bride
Kagan Publishing • 1 (800) 933-2667 • www.KaganOnline.com

2 DEFINITIONS

7. Absolute value of a number

Examples	Counterexamples
7 is the absolute value of –7 because –7 is seven units away from 0.	–7 is not the absolute value of –7.
4 is the absolute value of 4 because 4 is four units away from 0.	–3 is not the absolute value of 3.
5.3 is the absolute value of 5.3 because 5.3 is 5.3 units away from 0.	–8 is not the absolute value of –8.
9 is the absolute value of –9 because –9 is nine units away from 0.	–17 is not the absolute value of 17.

ACTIVITY

2 DEFINITIONS

8. Multiplicative inverse of a number

Examples	**Counterexamples**

$\frac{1}{3}$ is the multiplicative inverse of 3

because $\frac{1}{3} \times 3 = 1$.

$\frac{-4}{3}$ is the multiplicative inverse of $\frac{3}{4}$

because $\frac{-4}{3} \times \frac{-3}{4} = 1$.

$\frac{2}{5}$ is the multiplicative inverse of $\frac{5}{2}$

because $\frac{2}{5} \times \frac{5}{2} = 1$.

3 is not the multiplicative

inverse of 3 because $3 \times 3 = 9$.

$\frac{-4}{3}$ is not the multiplicative inverse

of $\frac{3}{4}$ because $\frac{-4}{3} \times \frac{3}{4} = -1$.

$\frac{2}{5}$ is not the multiplicative inverse

of $\frac{5}{1}$ because $\frac{2}{5} \times \frac{5}{1} = \frac{10}{5} = 2$.

8. Exponent

Examples	**Counterexamples**

3^2; 2 is an exponent and 3^2
    means $3 \cdot 3$

5^3; 3 is an exponent and 5^3
    means $5 \cdot 5 \cdot 5$

$(2x)^4$; 4 is an exponent and $(2x)^4$
    means $2x \cdot 2x \cdot 2x \cdot 2x$

4^3; 4 is not an exponent

5^3; 5 is not an exponent

x^4; x is not an exponent

Cooperative Learning and Algebra 1: Becky Bride
Kagan Publishing • 1 (800) 933-2667 • www.KaganOnline.com

3 NAME MY SETS

Structure: RallyRobin

Name all the sets to which each number belongs.

1. 0

2. $\dfrac{16}{3}$

3. $\dfrac{-8}{4}$

4. 8

5. 61

6. $-\sqrt{25}$

7. $\sqrt{49}$

8. 5.3

Answers:
1. whole, integer, rational
2. rational
3. integer, rational
4. natural, whole, integer, rational
5. natural, whole, integer, rational
6. integer, rational
7. natural, whole, integer, rational
8. rational

Cooperative Learning and Algebra 1: Becky Bride
Kagan Publishing • 1 (800) 933-2667 • www.KaganOnline.com

4 **NAME A NUMBER**

Structure: RallyRobin

1. Name a whole number that is not a natural number.

2. Name a rational number that is not an integer.

3. Name an integer that is not a whole number.

4. Name a rational number that is not a whole number.

Possible Answers:
1. 0
2. $\frac{2}{3}$
3. −5
4. −7

Cooperative Learning and Algebra 1: Becky Bride
Kagan Publishing • 1 (800) 933-2667 • www.KaganOnline.com

ACTIVITY

5 **INTEGER CARDS**

RATIONAL NUMBERS -2/3

Structure: Quiz-Quiz-Trade

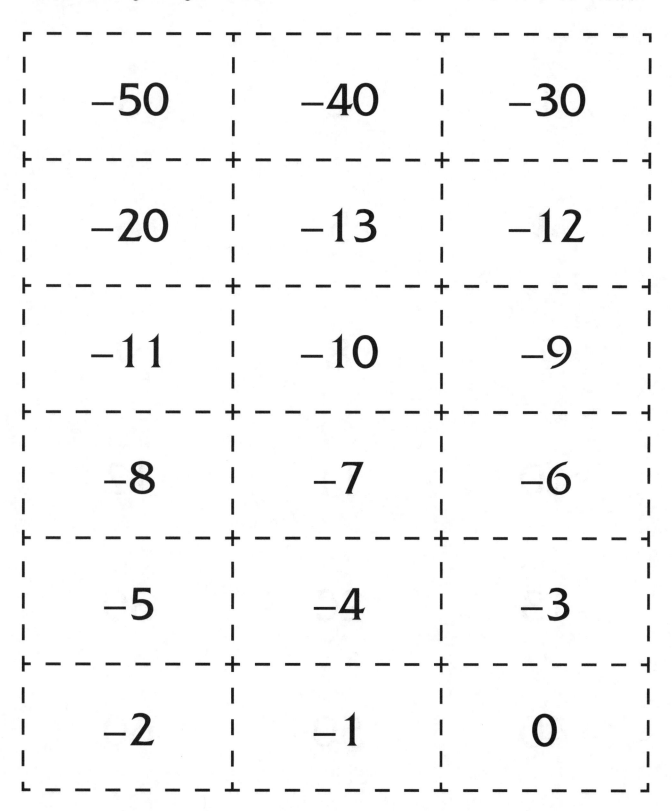

–50	–40	–30
–20	–13	–12
–11	–10	–9
–8	–7	–6
–5	–4	–3
–2	–1	0

ACTIVITY
5 **INTEGER CARDS**

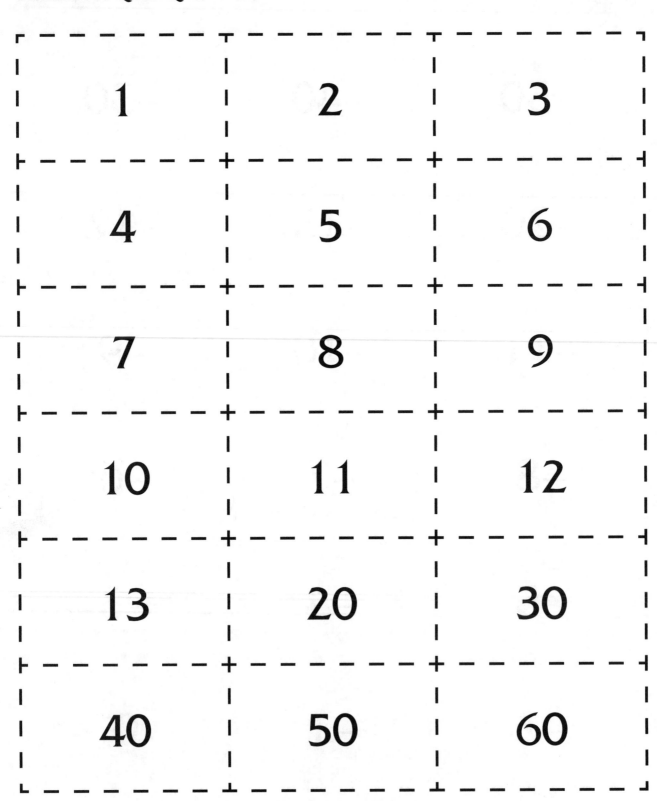

Structure: Quiz-Quiz-Trade

1	2	3
4	5	6
7	8	9
10	11	12
13	20	30
40	50	60

Cooperative Learning and Algebra 1: Becky Bride
Kagan Publishing • 1 (800) 933-2667 • www.KaganOnline.com

RATIONAL NUMBER CARDS

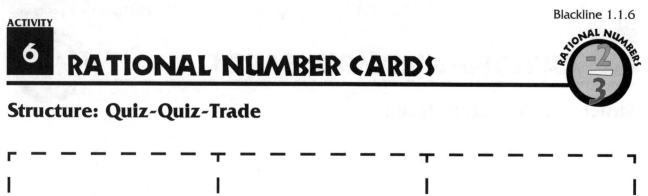

Structure: Quiz-Quiz-Trade

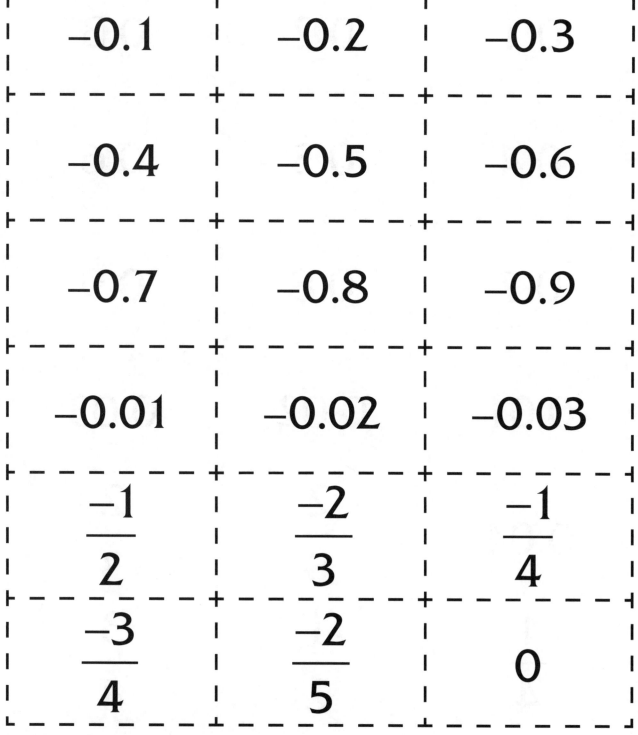

−0.1	−0.2	−0.3
−0.4	−0.5	−0.6
−0.7	−0.8	−0.9
−0.01	−0.02	−0.03
$\dfrac{-1}{2}$	$\dfrac{-2}{3}$	$\dfrac{-1}{4}$
$\dfrac{-3}{4}$	$\dfrac{-2}{5}$	0

ACTIVITY 6 RATIONAL NUMBER CARDS

Structure: Quiz-Quiz-Trade

0.1	0.2	0.3
0.4	0.5	0.6
0.7	0.8	0.9
0.01	0.02	0.03
0.04	$\dfrac{2}{5}$	$\dfrac{2}{3}$
$\dfrac{1}{4}$	$\dfrac{1}{2}$	$\dfrac{3}{4}$

Cooperative Learning and Algebra 1: Becky Bride
Kagan Publishing • 1 (800) 933-2667 • www.KaganOnline.com

LESSON 2
OPERATION ON INTEGERS

This lesson focuses on addition, subtraction, multiplication, and division of integers. Addition, subtraction, and multiplication have exploratory exercises using algebra tiles so the students will discover the "rules" of adding, subtracting, and multiplying integers. An exploratory activity for division is not included because those rules can be derived from fact families (rewriting multiplication problems as division problems). For those new to algebra tiles, an explanation of how they are used is included with each investigation. Following each investigation are two or more activities to process the operation just learned. The first processing is done using the algebra tiles and the second processing is done using the rules established in the investigation. The lesson ends with applications and a graphic organizer.

ACTIVITY
1 EXPLORING ADDITION OF INTEGERS

Working with algebra tiles for addition
Copy blackline master 1.2.1 on two different colors of cardstock so each student will have all the pages in each color. One color represents positive and the other color represents negative. These pages contain all the algebra tiles you will need for the activities in this chapter. The ones needed for this investigation are the small squares at the top of the first page. The remainder of the tiles on this page are used in later chapters. The length of each side of the small squares is one unit so that the area of the small squares is 1 square unit. Thus each small square is the representation of 1.

The fundamental concept required for all the investigations is that one positive square (say green) and one negative square (say red) together are equivalent to zero. The students must understand this first.

Take the problem 2 + 4. Students will place 2 positive squares (each representing 1 sq. unit) and 4 positive squares (each representing 1 sq. unit) on their workspace, put them together and will have 6 green squares or tiles. Since the tiles are green the answer is positive 6.

Adding two negative numbers is similar. Take the problem –1 + –3. Students will place one red tile and three more red tiles (each representing 1 sq. unit) in their workspace. Putting them together gives 4 red tiles so the answer is –4.

Adding a positive and a negative number works with the zero concept of one red tile and one green tile being equivalent to zero (have a dollar but owe a dollar so my balance is really zero). Take the problem – 3 + 5. Students will place three red tiles and five green tiles in their workspace. Then students pair a green tile with a red tile until no more pairs can be made to locate all the zeros. Removing the zeros from the workspace leaves two green tiles so the answer is positive 2.

After grouping the zeros together it looks like this.

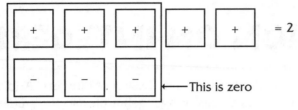

This is zero

Modeling with tiles is extremely important. The lower the academic ability of the child, the more modeling is necessary. The concept of one positive and one negative tile being equivalent to zero is important for all future integer investigations. Time spent here on this concept will save time later.

Problem 27 of the investigation asks the students to rewrite each problem above as an arithmetic expression that illustrates how you used the algebra tiles to get each answer. The purpose of this step is to bridge the concrete level with the abstract form of using numbers. So $6 + (-2)$ would look like: $4 + 2 + (-2)$. The students can see that the 2 and -2 will form the zero.

▶ Structure

• Solo-Pair Consensus-Team Consensus

▶ Materials

• 1 Blackline 1.2.2 per student
• 1 set of algebra tiles per student (Blackline 1.2.1)
• 1 pencil per student

Concrete

Solo

1. Have each student complete the investigation individually using the algebra tiles.

Pair Consensus

2. For each problem on the investigation, each student shares with his/her partner, using RallyRobin, his/her response. They discuss the problems on which they disagree, trying to come to consensus on the correct response. For the problems they can't reach consensus on, they mark these so they can focus on them during the team phase. Encourage the students to add to their responses if their partner verbalized an understanding they did not see.

Team Consensus

3. Each pair shares their responses, using RallyRobin, with the other pair in their team, augmenting their responses if necessary. When the teams are through sharing, each student should have a detailed, complete summary.

Cooperative Learning and Algebra 1: Becky Bride
Kagan Publishing • 1 (800) 933-2667 • www.KaganOnline.com

2 ADDITION USING TILES

Concrete

Setup:
In pairs, Student A is the Sage; Student B is the Scribe. Students fold a sheet of paper in half and each writes his/her name on one half.

1. The Sage, using algebra tiles, gives the Scribe step-by-step instructions on how to solve problem 1 using algebra tiles.

2. The Scribe records the Sage's solution step-by-step by drawing a diagram of the algebra tiles.

3. If the Sage is correct, the Scribe praises the Sage. Otherwise, the Scribe coaches, then praises.

4. Students switch roles for the next problem.

▶ **Structure**
• Sage-N-Scribe

▶ **Materials**
• Transparency 1.2.3
• 1 sheet of paper and pencil per pair
• 1 set of algebra tiles per pair

3 ADDITION USING RULES

Concrete

Setup:
In pairs, Student A is the Sage; Student B is the Scribe. Students fold a sheet of paper in half and each writes his/her name on one half.

1. The Sage gives the Scribe step-by-step instructions on how to do problem 1 using the rules of adding positive and negative numbers established in the investigation.

2. The Scribe records the Sage's solution step-by-step in writing.

3. If the Sage is correct, the Scribe praises the Sage. Otherwise, the Scribe coaches, then praises.

4. Students switch roles for the next problem.

▶ **Structure**
• Sage-N-Scribe

▶ **Materials**
• Transparency 1.2.3
• 1 sheet of paper and pencil per pair

Chapter 1: Working with Rational Numbers

Lesson Two

ACTIVITY 4

ADDITION PRACTICE

▶ **Structure**
· Mix-Freeze-Group

▶ **Materials**
· Set of integer cards or rational number cards (Blacklines 1.1.5 and/or 1.1.6)

Numeric

Setup:
Teacher prepares questions to which the answers are the number 3. Teacher distributes one card per student.

1. Students mix around the room.

2. Teacher calls "Freeze," and students freeze.

3. Teacher asks a question to which the answer is 3. (What is the sum of −3 and 6?)

4. Students group according to the number, and kneel down.

5. In their groups, Student 1 and Student 2 show their cards to Student 3 and ask Student 3 what the sum is.

6. Student 3 responds.

7. The two students coach and/or praise Student 3.

8. Student 2 and Student 3 show their cards to Student 1 and ask Student 1 what the sum is.

9. Student 1 responds.

10. The two students coach and/or praise Student 1.

11. Student 1 and Student 3 show their cards to Student 2 and ask Student 2 what the sum is.

12. Student 2 responds.

13. The two students coach and/or praise Student 2.

14. Students trade cards and politely say goodbye.

15. Repeat steps 1-14 as many times as you want the students to practice.

Management tip: To start and stop the mixing, music works well. When the music starts, students mix. When the music stops, the students freeze.

ACTIVITY 5

ADDING WITH MULTIPLE ADDENDS

▶ **Structure**
· Sage-N-Scribe

▶ **Materials**
· Transparency 1.2.4
· 1 sheet of paper and pencil per pair

Numeric

Setup:
In pairs, Student A is the Sage; Student B is the Scribe. Students fold a sheet of paper in half and each writes his/her name on one half.

1. The Sage gives the Scribe step-by-step instructions on how to solve problem 1 using the rules of adding positive and negative numbers.

2. The Scribe records the Sage's solution step-by-step in writing.

3. If the Sage is correct, the Scribe praises the Sage. Otherwise, the Scribe coaches, then praises.

4. Students switch roles for the next problem.

Cooperative Learning and Algebra 1: Becky Bride
Kagan Publishing • 1 (800) 933-2667 • www.KaganOnline.com

6 EXPLORING SUBTRACTION OF INTEGERS

Working with algebra tiles for subtraction

The students will use the same algebra tiles they used for the addition investigation—the 1 square unit. In addition to the zero concept, the students need to understand that the subtraction symbol means to remove. So the problem 4 – 5 means to remove 5 positive tiles from the 4 positive tiles. The zero concept plays a critical role in these problems. Again the success to this investigation is ample modeling with the tiles. Students must be comfortable with introducing zeros into the workspace and removing the appropriate tiles.

The first two parts of the investigation work with two positive numbers or two negative numbers. As an example, take the problem 7 – 3. The students would place 7 positive (green) tiles on their workspace. The students then need to remove 3 positive (green) tiles. This is easy for the students to do. The investigation then leads them into comparing each subtraction problem with its equivalent addition problem. The purpose is to begin to have the students see that subtracting a number is equivalent to adding the opposite of that number.

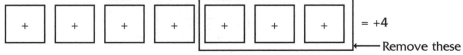

Subtracting two negative numbers works the same way. Take the problem (–5) – (–2).

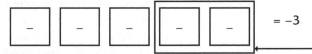

Subtracting numbers with unlike signs or numbers with like signs where the smaller of the two numbers appears first in the problem requires introducing zeros into the problem. For the problem 6 – (–2) the students begin by placing 6 positive tiles (green) on their workplace. Since all the tiles are positive, two negative tiles cannot be removed. So the students will have to introduce two zeros (1 green and 1 red) into the problem. The value of the problem has not changed because 6 + 0 + 0 = 6. Now the students can remove 2 negative (red) tiles. This leaves 8 positive (green) tiles so the answer is positive 8. This is illustrated below.

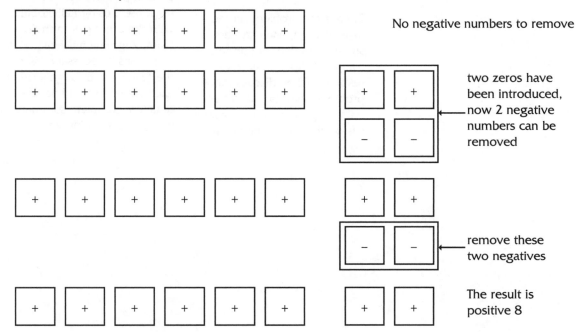

When the smaller number comes first, the students will have to introduce zeros on the workspace. Take the problem 5 – 8. The student can remove some positive tiles but not enough to complete the subtraction. The student will have to introduce 3 zeros into the problem so there will be 8 positive tiles to remove.

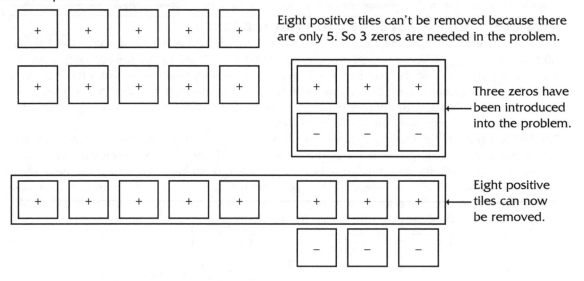

Eight positive tiles can't be removed because there are only 5. So 3 zeros are needed in the problem.

Three zeros have been introduced into the problem.

Eight positive tiles can now be removed.

▶ Structure
• Solo-Pair Consensus-Team Consensus

▶ Materials
• 1 Blackline 1.2.5 per student
• 1 set of algebra tiles per student (Blackline 1.2.1)
• 1 pencil per student

Concrete

Solo

1. Have each student complete the investigation individually using the algebra tiles. This could be done as homework.

Pair Consensus

2. For each problem on the investigation, each student shares with his/her partner, using RallyRobin, his/her response. They discuss the problems that they disagree on, trying to come to consensus on the correct response.

For the problems they can't reach consensus on, they mark these so they can focus on them during the team phase. Encourage the students to add to their responses if their partner verbalized an understanding they did not see.

Team Consensus

3. Each pair shares their responses, using RallyRobin, with the other pair in their team, augmenting their responses if necessary. When the teams are through sharing, each student should have a detailed, complete summary of how to subtract integers.

ACTIVITY

7 SUBTRACTION USING TILES

Concrete

Setup:
In pairs, Student A is the Sage; Student B is the Scribe. Students fold a sheet of paper in half and each writes his/her name on one half.

1. The Sage, using algebra tiles, gives the Scribe step-by-step instructions on how to solve problem 1 using algebra tiles.

2. The Scribe records the Sage's solution step-by-step by drawing a diagram of the algebra tiles.

3. If the Sage is correct, the Scribe praises the Sage. Otherwise, the Scribe coaches, then praises.

4. Students switch roles for the next problem.

▶ **Structure**
• Sage-N-Scribe

▶ **Materials**
• Transparency 1.2.6
• 1 sheet of paper and pencil per pair
• 1 set of algebra tiles per pair

ACTIVITY

8 SUBTRACTION USING RULES

Numeric

Setup:
In pairs, Student A is the Sage; Student B is the Scribe. Students fold a sheet of paper in half and each writes his/her name on one half.

1. The Sage gives the Scribe step-by-step instructions on how to do problem 1 using the rules of subtracting positive and negative numbers established in the investigation.

2. The Scribe records the Sage's solution step-by-step in writing.

3. If the Sage is correct, the Scribe praises the Sage. Otherwise, the Scribe coaches, then praises.

4. Students switch roles for the next problem.

Note: At first, you may want to have the students rewrite the problems, changing subtraction to addition and the number following to its opposite, especially for a 2-year Algebra 1 class. Then the students could go back and finish the problems if they mastered the rewriting. Otherwise, more practice with rewriting may be necessary.

▶ **Structure**
• Sage-N-Scribe

▶ **Materials**
• Transparency 1.2.6
• 1 sheet of paper and pencil per pair

ACTIVITY

9 SUBTRACTION PRACTICE

▶ **Structure**
• Mix-Freeze-Group

▶ **Materials**
• Set of integer cards or rational number cards (Blacklines 1.1.5 and/or 1.1.6)

Numeric

Setup:
Teacher prepares questions to which the answers are the number 3. Teacher distributes one card per student.

1. Students mix around the room.

2. Teacher calls "Freeze," and students freeze.

3. Teacher asks a question to which the answer is 3. (What is the difference of –2 and 5?)

4. Students group according to the number, and kneel down.

5. In their groups, Student 1 and Student 2 show their cards to Student 3 and ask Student 3 what the difference is.

6. Student 3 responds.

7. The two students coach and/or praise Student 3.

8. Student 2 and Student 3 show their cards to Student 1 and ask Student 1 what the difference is.

9. Student 1 responds.

10. The two students coach and/or praise Student 1.

11. Student 1 and Student 3 show their cards to Student 2 and ask Student 2 what the difference is.

12. Student 2 responds.

13. The two students coach and/or praise Student 2.

14. Students trade cards and politely say goodbye.

15. Repeat steps 1-14 as many times as you want the students to practice.

Management tip: To start and stop the mixing, music works well. When the music starts, students mix. When the music stops, the students freeze.

ACTIVITY

10 CAN YOU HANDLE THE MIXTURE?

▶ **Structure**
• Sage-N-Scribe

▶ **Materials**
• Transparency 1.2.7
• 1 sheet of paper and pencil per pair

Numeric

Setup:
In pairs, Student A is the Sage; Student B is the Scribe. Students fold a sheet of paper in half and each writes his/her name on one half.

1. The Sage gives the Scribe step-by-step instructions on how to solve problem 1 using the rules of adding positive and negative numbers.

2. The Scribe records the Sage's solution step-by-step in writing.

3. If the Sage is correct, the Scribe praises the Sage. Otherwise, the Scribe coaches, then praises.

4. Students switch roles for the next problem.

Cooperative Learning and Algebra 1: Becky Bride
Kagan Publishing • 1 (800) 933-2667 • www.KaganOnline.com

ACTIVITY

11 EXPLORING MULTIPLICATION OF INTEGERS

Working with algebra tiles for multiplication

The students will be using the same algebra tiles used for the previous two investigations—1 square unit tiles. Reviewing the concept of multiplication is essential. The problem $2 \cdot 3$ means 2 groups of 3 items. The first number will indicate whether the students will put (positive) groups of tiles on their workspace, or whether they need to remove (negative) groups of tiles from their workspace. Multiplication is illustrated in the order of multiplying two positive numbers, multiplying a positive number with a negative number, multiplying a negative number with a positive number, and finally multiplying two negative numbers.

To multiply two positive numbers, the problem $2 \cdot 3$ will be used. The positive 2 tells the students to put two groups of three green tiles (positive) on their workspace. Once done, the students will see that there are now 6 green tiles (positive).

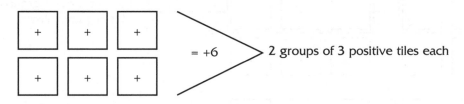

$= +6$ > 2 groups of 3 positive tiles each

In the problem $2 \cdot (-3)$, the positive 2 tells the students to put two groups of three red tiles (negative) on their workspace. The students will easily see that it is equal to -6.

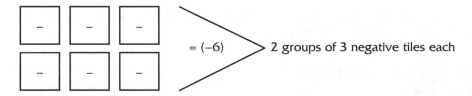

$= (-6)$ > 2 groups of 3 negative tiles each

In the problem $(-)2 \cdot 3$ the negative 2 tells the students to remove two groups of three green tiles (positive) from their workspace. There are no tiles in their workspace so the students must place six zeros on their workspace. They had no tiles to begin with and because they added zeros they technically still have nothing of value in their workspace. The students will then be able to remove 2 groups of 3 green (positive) tiles, leaving 6 negative tiles.

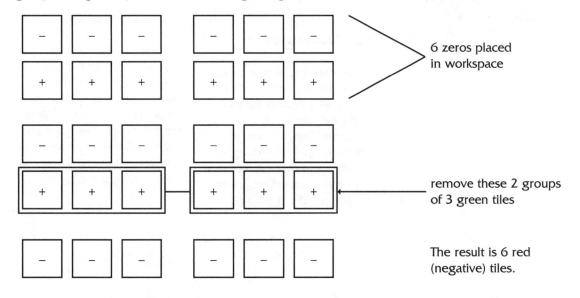

6 zeros placed in workspace

remove these 2 groups of 3 green tiles

The result is 6 red (negative) tiles.

Chapter 1: Working with Rational Numbers

Lesson Two

To multiply a negative number with a negative number, the students must understand the previous problem. The students will have no tiles in their workspace, and the first negative number will tell them to remove tiles (which aren't there), thus requiring the placement of zeros.

In the problem (–)2 · (–3) the negative 2 tells the students to remove two groups of three red tiles (negative) from their workspace. There are no tiles in their workspace, so the students must place six zeros on their workspace. They had no tiles to begin with and because they added zeros they technically still have nothing of value in their workspace. The students will then be able to remove 2 groups of 3 red (negative) tiles, leaving 6 positive tiles.

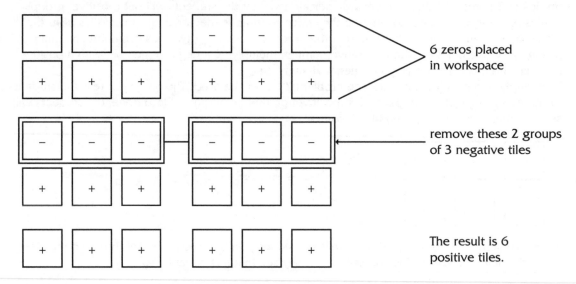

6 zeros placed in workspace

remove these 2 groups of 3 negative tiles

The result is 6 positive tiles.

The success of this investigation depends on ample modeling using the tiles. If the students are confused, then you need to model with the tiles more.

▶ **Structure**
• Solo-Pair Consensus-Team Consensus

▶ **Materials**
• 1 Blackline 1.2.8 per student
• 1 set of algebra tiles per student (Blackline 1.2.1)
• 1 pencil per student

Concrete

Solo

1. Have each student complete the investigation individually using the algebra tiles.

Pair Consensus

2. For each problem on the investigation, each student shares with his/her partner, using RallyRobin, his/her response. They discuss the problems that they disagree on, trying to come to consensus on the correct response. For the problems they can't reach consensus on, they mark these so they can focus on them during the team phase. Encourage the students to add to their responses if their partner verbalized an understanding they did not see.

Team Consensus

3. Each pair shares their responses, using RallyRobin, with the other pair in their team, augmenting their responses if necessary. When the teams are through sharing, each student should have a detailed, complete summary of how to multiply integers.

Cooperative Learning and Algebra 1: Becky Bride
Kagan Publishing • 1 (800) 933-2667 • www.KaganOnline.com

12 MULTIPLICATION USING TILES

Concrete

Setup:
In pairs, Student A is the Sage; Student B is the Scribe. Students fold a sheet of paper in half and each writes his/her name on one half.

1. The Sage, using algebra tiles, gives the Scribe step-by-step instructions on how to solve problem 1 using algebra tiles.

2. The Scribe records the Sage's solution step-by-step by drawing a diagram of the algebra tiles.

3. If the Sage is correct, the Scribe praises the Sage. Otherwise, the Scribe coaches, then praises.

4. Students switch roles for the next problem.

▶ **Structure**
• Sage-N-Scribe

▶ **Materials**
• Transparency 1.2.9
• 1 sheet of paper and pencil per pair
• 1 set of algebra tiles per pair

13 MULTIPLICATION USING RULES

Numeric

Setup:
In pairs, Student A is the Sage; Student B is the Scribe. Students fold a sheet of paper in half and each writes his/her name on one half.

1. The Sage, using algebra tiles, gives the Scribe step-by-step instructions using the rules of multiplying positive and negative numbers.

2. The Scribe records the Sage's solution step-by-step by writing.

3. If the Sage is correct, the Scribe praises the Sage. Otherwise, the Scribe coaches, then praises.

4. Students switch roles for the next problem.

▶ **Structure**
• Sage-N-Scribe

▶ **Materials**
• Transparency 1.2.9
• 1 sheet of paper and pencil per pair

Chapter 1: Working with Rational Numbers

Lesson Two

14 PROCESSING MULTIPLICATION AND DIVISION

▶ **Structure**
• Sage-N-Scribe

▶ **Materials**
• Transparency 1.2.10
• 1 sheet of paper and pencil per pair

Numeric

Setup:
In pairs, Student A is the Sage; Student B is the Scribe. Students fold a sheet of paper in half and each writes his/her name on one half.

1. The Sage gives the Scribe step-by-step instructions on how to solve problem 1 using the rules of multiplying and dividing positive and negative numbers.

2. The Scribe records the Sage's solution step-by-step in writing.

3. If the Sage is correct, the Scribe praises the Sage. Otherwise, the Scribe coaches, then praises.

4. Students switch roles for the next problem.

15 OPERATIONS ON INTEGERS

▶ **Structure**
• Simultaneous RoundTable

▶ **Materials**
• Transparency 1.2.11
• 1 sheet of paper and pencil per student

Numeric

Setup:
• Teammate 1 writes the two numbers in problem 1 at the top of his/her paper.
• Teammate 2 writes the two numbers in problem 2 at the top of his/her paper.
• Teammate 3 writes the two numbers in problem 3 at the top of his/her paper.
• Teammate 4 writes the two numbers in problem 4 at the top of his/her paper.

Round One

1. All four students respond simultaneously by adding the numbers found at the top of the paper and initialing his/her work.

2. Teacher signals time.

3. Students pass papers one person clockwise. Each teammate checks the addition problem on the paper he/she received, coaching the teammate who did the problem if it is incorrect. Then all four students respond simultaneously by subtracting the numbers at the top of the paper and initialing his/her work.

4. Teacher signals time.

5. Students pass their papers one person clockwise. Each teammate checks the subtraction problem on the paper he/she received, coaching the teammate who did the problem if it is incorrect. Then all four students respond simultaneously by multiplying the numbers at the top of the

paper and initialing his/her work.

6. Teacher signals time.

7. Students pass their papers one person clockwise. Each teammate checks the multiplication problem on the paper he/she received, coaching the teammate who did the problem if it is incorrect. Then all four students respond simultaneously by dividing the numbers at the top of the paper and initialing his/her work.

8. Students pass their papers one person clockwise. Each teammate checks the division problem on the paper he/she received, coaching the teammate who did the problem if it is incorrect.

Rounds 2-3

Repeat steps 1-8 for the remaining problems.

Cooperative Learning and Algebra 1: Becky Bride
Kagan Publishing • 1 (800) 933-2667 • www.KaganOnline.com

16 OPERATIONS ON RATIONAL NUMBERS

Numeric

Setup:
- Teammate 1 writes the two numbers in problem 1 at the top of his/her paper.
- Teammate 2 writes the two 2 the top of his/her paper.
- Teammate 3 writes the two numbers in problem 3 at the top of his/her paper.
- Teammate 4 writes the two numbers in problem 4 at the top of his/her paper.

Round One

1. All four students respond simultaneously by adding the numbers found at the top of the paper and initialing his/her work.

2. Teacher signals time.

3. Students pass papers one person clockwise. Each teammate checks the addition problem on the paper he/she

received, coaching the teammate who did the problem if it is incorrect. Then all four students respond simultaneously by subtracting the numbers at the top of the paper and initialing his/her work.

4. Teacher signals time.

5. Students pass their papers one person clockwise. Each teammate checks the subtraction problem on the paper he/she received, coaching the teammate who did the problem if it is incorrect. Then all four students respond simultaneously by multiplying the numbers at the top of the paper and initialing his/her work.

6. Teacher signals time.

7. Students pass their papers one person clockwise. Each teammate checks the multiplication problem on the paper

▶ **Structure**
- Simultaneous RoundTable

▶ **Materials**
- Transparency 1.2.12
- 1 sheet of paper and pencil per student

he/she received, coaching the teammate who did the problem if it is incorrect. Then all four students respond simultaneously by dividing the numbers at the top of the paper and initialing his/her work.

8. Students pass their papers one person clockwise. Each teammate checks the division problem on the paper he/she received, coaching the teammate who did the problem if it is incorrect.

Rounds 2-3

Repeat steps 1-8 for the remaining problems.

17 OPERATIONS ON INTEGERS IN THE REAL WORLD

Connecting

Setup:
In pairs, Student A is the Sage; Student B is the Scribe. Students fold a sheet of paper in half and each writes his/her name on one half.

1. The Sage gives the Scribe step-by-step instructions on how to solve problem 1.

2. The Scribe records the Sage's solution step-by-step in writing.

3. If the Sage is correct, the Scribe praises the Sage. Otherwise, the Scribe coaches, then praises.

4. Students switch roles for the next problem.

▶ **Structure**
- Sage-N-Scribe

▶ **Materials**
- 1 Blackline 1.2.13 per pair
- 1 sheet of paper and pencil per pair

Note: Make sure the students use negative numbers for loan balances, bank withdrawals, or credit card balances.

WHAT DID WE LEARN?

▶ **Structure**
• RoundTable Consensus

▶ **Materials**
• 1 large sheet of paper per team
• different color pen or pencil for each student on the team

Synthesis via Graphic Organizer

1. Each teammate signs his/her name in the upper-right corner of the team paper with the color pen/pencil he/she is using.

2. One teammate writes "Operations on Integers" in the center of the team paper in a rectangle.

3. Teammate 1 shares with the team one core concept he/she learned in the unit.

4. The student checks for consensus.

5. The teammates show agreement or lack of agreement with thumbs up or down.

6. If there is agreement, the students celebrate and the teammate records the core concept on the graphic organizer, connecting it with a line to the main idea, "Operations on Integers." If not, teammates discuss the response until there is agreement and then they celebrate.

7. Play continues with the next student's core concept until all core concepts are exhausted.

8. Repeat steps 3-7, with teammates adding details to each core concept and making bridges between related ideas.

Cooperative Learning and Algebra 1: Becky Bride
Kagan Publishing • 1 (800) 933-2667 • www.KaganOnline.com

ACTIVITY

1 ALGEBRA TILES

1, x, and x^2

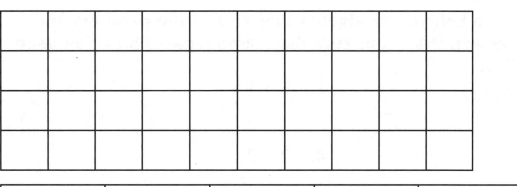

RATIONAL NUMBERS $\frac{-2}{3}$

ACTIVITY 1
EXPLORING ADDITION OF INTEGERS

For each problem below, use algebra tiles to add the numbers and record your answer. Draw a picture that demonstrates how you used the algebra tiles.

1. 3 + 4

2. 5 + 1

3. 6 + 3

4. 7 + 2

5. Compare the numbers in each problem with its answer. In order to add each pair of positive numbers above, do you add, subtract, multiply, or divide the numbers?

6. When adding two positive numbers, what is the sign of the answer?

7. Explain how to add a pair of positive numbers.

8. Explain a real-world situation that involves the addition of two positive numbers.

For each problem below, use algebra tiles to add the numbers and record your answer. Draw a picture that demonstrates how you used the algebra tiles.

9. −4 + (− 2)

10. −3 + (−5)

11. −1 + (−6)

12. −4 + (−7)

Cooperative Learning and Algebra 1: Becky Bride
Kagan Publishing • 1 (800) 933-2667 • www.KaganOnline.com

ACTIVITY
1

EXPLORING ADDITION
OF INTEGERS

13. Compare the numbers in each problem with its answer. In order to add each pair of negative numbers on page 34, do you add, subtract, multiply, or divide the numbers?

14. When adding two negative numbers, what is the sign of the answer?

15. Explain how to add a pair of negative numbers.

16. Explain a real-world situation that involves the addition of two negative numbers.

For each problem below, use algebra tiles to add the numbers and record your answer. Draw a picture that demonstrates how you used the algebra tiles.

17. $5 + (-3)$ 18. $-8 + 4$

19. $5 + (-7)$ 20. $-6 + 9$

21. $-5 + 1$ 22. $-2 + 6$

23. $6 + (-3)$ 24. $2 + (-8)$

ACTIVITY
1

EXPLORING ADDITION
OF INTEGERS

25. Compare the numbers in each problem with its answer. In order to add each pair of numbers on page 35, do you add, subtract, multiply, or divide the numbers?

26. Some answers were positive and some were negative. How do you determine the sign of the answer when adding two numbers with unlike (or different) signs?

27. Using numbers, rewrite each problem on pages 34-35 as an arithmetic expression that illustrates how you used the algebra tiles to arrive at each answer.

28. Explain a real-world situation that involves the addition of two numbers with unlike signs.

29. Summarize the entire investigation, describing how to add numbers with like signs and how to add numbers with unlike signs.

Cooperative Learning and Algebra 1: Becky Bride
Kagan Publishing • 1 (800) 933-2667 • www.KaganOnline.com

RATIONAL NUMBERS

$\frac{-2}{3}$

ACTIVITIES 2&3
ADDITION USING TILES
ADDITION USING RULES

Structure: Sage-N-Scribe

For each problem below, a) perform the indicated operation and b) classify each answer with the set(s) of numbers to which it belongs.

1. $-4 + 8$

2. $8 + (-3)$

3. $6 + (-10)$

4. $-12 + 9$

5. $-5 + (-8)$

6. $4 + 7$

7. $-9 + 9$

8. $10 + (-3)$

9. $-3 + (-7)$

10. $9 + 2$

Answers:
1. 4; natural, whole, integer, rational
2. 5; natural, whole, integer, rational
3. −4; integer, rational
4. −3; integer, rational
5. −13; integer, rational
6. 11; natural, whole, integer, rational
7. 0; whole, integer, rational
8. 7; natural, whole, integer, rational
9. −10; integer, rational
10. 11; natural, whole, integer, rational

ACTIVITY
5

ADDING WITH MULTIPLE ADDENDS

Structure: Sage-N-Scribe

Perform the indicated operation.

1. 34 + (−23) + 62

2. 18 + (−54) + (−2)

3. −3 + (−19) + 49

4. −78 + (−22) + (−53)

5. 34 + 86 + 21

6. 43 + (−12) + (−73)

Answers:
1. 73
2. −38
3. 27
4. −153
5. 141
6. −42

Cooperative Learning and Algebra 1: Becky Bride
Kagan Publishing • 1 (800) 933-2667 • www.KaganOnline.com

ACTIVITY 6

EXPLORING SUBTRACTION OF INTEGERS

RATIONAL NUMBERS $\frac{-2}{3}$

For each problem below, use algebra tiles to subtract the numbers; then record your answer. Draw a picture that demonstrates how you used the algebra tiles.

1. 6 – 4

2. 5 – 2

3. 6 – 1

4. 8 – 3

5. Compute 6 + (–4). Is this similar to problem 1? If so, in what way?

6. Compute 5 + (–2). Is this similar to problem 2? If so, in what way?

7. Compute 6 + (–1). Is this similar to problem 3? If so, in what way?

8. Compute 8 + (–3). Is this similar to problem 4? If so, in what way?

9. The above examples demonstrate that a subtraction problem can be rewritten as an addition problem. Rewrite the following subtraction problems into equivalent addition problems; then follow the rules for addition to compute the answers.

   a) 17 – 4

   b) 15 – 12

   c) 23 – 9

   d) 32 – 18

ACTIVITY 6
EXPLORING SUBTRACTION OF INTEGERS

RATIONAL NUMBERS
$\frac{-2}{3}$

10. A neighborhood friend has an assignment where she has to rewrite her subtraction problems as addition problems and does not understand it. Write below, in your own words, how you would explain how to transform a subtraction problem into an addition problem.

For each problem below, use algebra tiles to subtract the numbers; then record your answer. Draw a picture that demonstrates how you used the algebra tiles.

11. $-4 - (-2)$

12. $-3 - (-6)$

13. $-1 - (-7)$

14. $-5 - (-4)$

15. Compute $-4 + 2$. Is this similar to problem 11? If so, in what way?

16. Compute $-3 + 6$. Is this similar to problem 12? If so, in what way?

17. Compute $-1 + 7$. Is this similar to problem 13? If so, in what way?

18. Compute $-5 + 4$. Is this similar to problem 14? If so, in what way?

Cooperative Learning and Algebra 1: Becky Bride
Kagan Publishing • 1 (800) 933-2667 • www.KaganOnline.com

EXPLORING SUBTRACTION OF INTEGERS

19. The previous examples demonstrate that a subtraction problem can be rewritten as an addition problem. Rewrite the following subtraction problems into equivalent addition problems; then follow the rules for addition to compute the answers.

    a) $-13 - (-4)$

    b) $-16 - (-12)$

    c) $-22 - (-9)$

    d) $-36 - (-18)$

20. Your parent is looking over your shoulder and notices that you rewrite each subtraction problem as an addition problem. Your parent doesn't understand. Write below, in your own words, what you say to explain it.

For each problem below, use algebra tiles to subtract the numbers; then record your answer. Draw a picture that demonstrates how you used the algebra tiles.

21. $5 - (-3)$

22. $-8 - 4$

23. $5 - (-6)$

24. $-6 - 9$

25. Compute $5 + 3$. Is this similar to problem 21? If so, in what way?

EXPLORING SUBTRACTION OF INTEGERS

26. Compute – 8 + (–4). Is this similar to problem 22? If so, in what way?

27. Compute 5 + 6. Is this similar to problem 23? If so, in what way?

28. Compute –6 + (–9). Is this similar to problem 24? If so, in what way?

29. The above examples demonstrate that a subtraction problem can be rewritten as an addition problem. Rewrite the following subtraction problems into equivalent addition problems; then follow the rules for addition to compute the answers.

   a)   –13 – 4   b)   16 – (–12)

   c)   2 – (–29)   d)   –11 – 18

30. Summarize the entire investigation by writing, in your own words, how to subtract positive and negative numbers.

Cooperative Learning and Algebra 1: Becky Bride
Kagan Publishing • 1 (800) 933-2667 • www.KaganOnline.com

SUBTRACTION USING TILES
SUBTRACTION USING RULES

RATIONAL NUMBERS
$\frac{-2}{3}$

Structure: Sage-N-Scribe

Perform the indicated operation.

1. $5 - (-3)$ 2. $4 - 9$

3. $-3 - 5$ 4. $2 - 6$

5. $-6 - (-8)$ 6. $12 - (-4)$

7. $7 - 3$ 8. $-9 - (-4)$

Answers:
1. 8
2. −5
3. −8
4. −4
5. 2
6. 16
7. 4
8. −5

CAN YOU HANDLE THE MIXTURE?

Structure: Sage-N-Scribe

Perform the indicated operations.

1. –31 + 42 – 16

2. 29 – (–14) + (–54)

3. 73 + (–37) – (–24)

4. –82 – 44 + (–56)

5. 34 – (–19) + 75

6. 61 – 84 + (–55)

Answers:
1. –5
2. –11
3. 60
4. –182
5. 128
6. –78

Cooperative Learning and Algebra 1: Becky Bride
Kagan Publishing • 1 (800) 933-2667 • www.KaganOnline.com

RATIONAL NUMBERS $\frac{-2}{3}$

ACTIVITY 11 EXPLORING MULTIPLICATION OF INTEGERS

For each problem below, use algebra tiles to multiply the numbers; then record your answer. Draw a picture that demonstrates how you used the algebra tiles.

1. $6 \cdot 4$

2. $5 \cdot 2$

3. $4 \cdot 3$

4. $3 \cdot 2$

5. Compare the numbers in each problem to its answer. In order to multiply each pair of positive numbers above, did you add, subtract, multiply, or divide the numbers?

6. When multiplying two positive numbers, what is the sign of the answer?

7. Explain how this is similar to adding two positive numbers.

For each problem below, use algebra tiles to multiply the numbers; then record your answer. Draw a picture that demonstrates how you used the algebra tiles.

8. $5 \cdot (-3)$

9. $4 \cdot (-8)$

10. $5 \cdot (-7)$

11. $9 \cdot (-6)$

ACTIVITY 11 EXPLORING MULTIPLICATION OF INTEGERS

12. Compare the numbers in each problem to its answer. In order to multiply each pair of numbers on page 45, did you add, subtract, multiply, or divide the numbers?

13. When multiplying a positive number with a negative number, what is the sign of the answer?

14. Explain how this is different from adding two numbers with unlike signs.

For problems 15-18, use algebra tiles to multiply the numbers; then record your answers. Draw a picture to demonstrate how you used the algebra tiles.

15. $(-4)(3)$

16. $(-2)(3)$

17. $(-5)(3)$

18. $(-6)(2)$

19. Compare the numbers in each problem to its answer. In order to multiply each pair of numbers (with different signs) above, did you add, subtract, multiply, or divide the numbers?

20. When multiplying a negative number with a positive number, what is the sign of the answer?

Cooperative Learning and Algebra 1: Becky Bride
Kagan Publishing • 1 (800) 933-2667 • www.KaganOnline.com

EXPLORING MULTIPLICATION OF INTEGERS

21. How are the problems 8-11 and 15-18 similar?

For problems 22-25, use algebra tiles to multiply the numbers; then record your answers. Draw a diagram to demonstrate how you multiplied the numbers.

22. $(-4)(-3)$ 23. $(-2)(-3)$

24. $(-5)(-3)$ 25. $(-6)(-2)$

26. Compare the numbers in each problem to its answer. In order to multiply each pair of negative numbers above, did you add, subtract, multiply, or divide the numbers?

27. When multiplying 2 negative numbers, what is the sign of the answer?

28. How is multiplying 2 negative numbers different from adding 2 negative numbers?

29. Summarize the investigation of multiplying positive and negative numbers, two numbers at a time. Include in your summary multiplying two numbers with like signs and two numbers with unlike signs.

MULTIPLICATION USING TILES
MULTIPLICATION USING RULES

Structure: Sage-N-Scribe

Perform the indicated operation.

1. (3)(–4)

2. (–3)(–5)

3. (–2)(4)

4. (–3)(1)

5. (–4)(–4)

6. (–3)(6)

7. (9)(1)

8. (–5)(–2)

9. (–1)(7)

10. (8)(–2)

Answers:
1. –12
2. 15
3. –8
4. –3
5. 16
6. –18
7. 9
8. 10
9. –7
10. –16

Cooperative Learning and Algebra 1: Becky Bride
Kagan Publishing • 1 (800) 933-2667 • www.KaganOnline.com

ACTIVITY 14 — PROCESSING MULTIPLICATION AND DIVISION

RATIONAL NUMBERS $\frac{-2}{3}$

Structure: Sage-N-Scribe

Perform the indicated operation.

1. $(3)(-4)\,(1)$

2. $(-48) \div 12$

3. $(-12) \div (-2)$

4. $(-3)(1)(-3)$

5. $(-4)(-4)(-2)$

6. $24 \div (-8)$

7. $(-16) \div 4$

8. $(-1)(-7)(-3)$

9. $(8)(-2)(3)$

10. $(-40) \div (-8)$

11. $72 \div (-8)$

12. $(-5)(-2)(-3)$

Answers:
1. −12
2. −4
3. 6
4. 9
5. −32
6. −3
7. −4
8. −21
9. −48
10. 5
11. −9
12. −30

Cooperative Learning and Algebra 1: Becky Bride
Kagan Publishing • 1 (800) 933-2667 • www.KaganOnline.com

OPERATIONS ON INTEGERS

Structure: Simultaneous RoundTable

Round 1

1. 16 and (– 4)

2. –24 and (–2)

3. –40 and 8

4. –9 and (–3)

Round 2

1. –30 and 5

2. 10 and (–2)

3. –18 and (–3)

4. 8 and (–4)

Round 3

1. –15 and (–7)

2. 5 and (–3)

3. 10 and (–3)

4. –4 and 8

Cooperative Learning and Algebra 1: Becky Bride
Kagan Publishing • 1 (800) 933-2667 • www.KaganOnline.com

ACTIVITY 16 OPERATIONS ON RATIONAL NUMBERS

Structure: Simultaneous RoundTable

Round 1

1. 1.2 and (−5.6)

2. $\dfrac{-2}{3}$ and $\left(\dfrac{-4}{5}\right)$

3. $\dfrac{-5}{9}$ and $\dfrac{5}{6}$

4. −0.3 and (−2.5)

Round 2

1. $1\dfrac{1}{4}$ and $\left(\dfrac{-1}{8}\right)$

2. 3.6 and (−0.45)

3. −5.1 and (−0.4)

4. $\dfrac{-3}{4}$ and $2\dfrac{1}{5}$

Round 3

1. $\dfrac{-7}{9}$ and $\left(\dfrac{-3}{8}\right)$

2. $\dfrac{-3}{5}$ and $\left(\dfrac{3}{10}\right)$

3. 0.34 and (−.8)

4. −4.3 and 8.5

**OPERATIONS ON INTEGERS
IN THE REAL WORLD**

For each problem, write a numerical expression that models the situation and evaluate the expression.

1. The temperature on January 4 began at 5° below zero. The high temperature reached that day was 18°. What was the increase in temperature?

2. Kendra's checking account balance is $42.16. She writes a check for $57.63. What is her new balance?

3. Miguel's credit card bill arrives. The balance on the account is $126.13. He makes a $50 payment. What will his new balance be?

4. Last week Joni made five $20 ATM withdrawals from her savings account. What was the total amount withdrawn?

5. Henry's car loan has a balance of $5,283.27. If he makes a $498 payment, what is his new balance?

6. Sally was given a $50 gift card to the local mall for her birthday. If her purchases total $37.25 at Pac Sun, what is the balance on her gift card?

Answers:
1. $18 - (-5)$; 23
2. $42.16 - 57.63$; -15.47
3. $-126.13 + 50$; -76.13
4. $-20(5)$; -100
5. $-5283.27 + 498$; -4785.27
6. $50 + (-37.25)$; 12.75

Cooperative Learning and Algebra 1: Becky Bride
Kagan Publishing • 1 (800) 933-2667 • www.KaganOnline.com

LESSON 3
ORDER OF OPERATIONS

This lesson explores the order of operations in three exploratory activities. The first exploratory activity assumes the students know that the operations addition and subtraction are equal in priority, so they are performed as they appear from left to right. It also assumes that the students know multiplication and division are equal in priority, so they are done as they appear from left to right. The first exploration has students explore the priority of addition/subtraction with multiplication/division. The following two activities process the results of this exploration. Students will then investigate the priority of multiplication/division with exponents. This exploratory activity is followed by two activities that process the order of operations from exponents through addition/subtraction. The final investigation leads the students to explore the role of grouping symbols in the hierarchy of order of operations. This activity is also followed by two activities for the students to practice the order of operations. Finally, the lesson ends with an activity that synthesizes what the students have learned through the creation of a graphic organizer.

The order of operations is split into three investigations to allow students to master each segment of the concept. When the three pieces are put together, the students will have a better mastery of the order of operations, compared to giving them the entire chunk at once. Also this is conducive to the two-year algebra programs because those students need algebra in much smaller pieces than a one-year course. Thus, it is a win-win situation for all.

ACTIVITY 1
INVESTIGATING ADDITION/SUBTRACTION VS. MULTIPLICATION/DIVISION

Exploratory

Solo

1. Have each student complete the investigation individually using the scientific calculator. Make sure to inform the students to input the entire expression into his/her calculator before hitting the equal key.

Pair Consensus

2. For each problem on the investigation, each student shares with his/her partner, using RallyRobin, his/her response. They discuss the problems that they disagree on, trying to come to consensus on the correct response. For the problems they can't reach consensus on, they mark these so they can focus on them during the team phase. Encourage the students to add to their responses if their partner verbalized an understanding they did not see.

Team Consensus

3. Each pair shares their responses, using RallyRobin, with the other pair in their team, augmenting their responses if necessary. When the teams are through sharing, each student should have a detailed, complete summary.

▶ **Structure**
· Solo-Pair Consensus-Team Consensus

▶ **Materials**
· 1 Blackline 1.3.1 per student
· 1 scientific calculator per student
· 1 pencil per student

ACTIVITY

2 PROCESSING ADDITION/SUBTRACTION VS. MULTIPLICATION/DIVISION

▶ **Structure**
• Sage-N-Scribe

▶ **Materials**
• Transparency 1.3.2
• 1 sheet of paper and pencil per pair

Numeric

Setup:
In pairs, Student A is the Sage; Student B is the Scribe. Students fold a sheet of paper in half and each writes his/her name on one half.

1. The Sage gives the Scribe step-by-step instructions on how to solve problem 1.

2. The Scribe records the Sage's solution step-by-step in writing.

3. If the Sage is correct, the Scribe praises the Sage. Otherwise, the Scribe coaches, then praises.

4. Students switch roles for the next problem.

ACTIVITY

3 WRITE A PROBLEM

▶ **Structure**
• Placemat Consensus

▶ **Materials**
• Transparency 1.3.3
• 1 large sheet of paper (placemat) per team
• 1 pencil per student

Numeric Higher Level Thinking

1. Each team draws a rectangle in the center of the paper that will be the central team space and sections off the area around the large rectangle for individual team member space, similar to the diagram below. In the central team space, each team separates it into 3 sections, one for each problem.

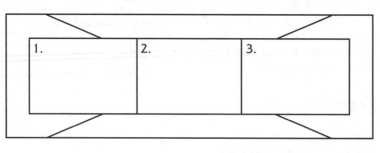

2. Teacher provides an answer and operations (see Transparency 1.3.3).

3. Teammates respond simultaneously in their individual space, writing as many problems as they can using the operations in the order given to get the given answer in the time allotted.

4. Teammate 1 shares with the team one item he/she has written.

5. Teammates discuss the item.

6. If there is consensus that the item is correct, teammate 1 records his/her problem in the team space of the placemat set aside for problem 1.

7. Repeat steps 4-6, so each teammate, in turn, suggests an idea and records the team consensus.

8. Repeat steps 2-7 for the remaining problems.

4 INVESTIGATING MULTIPLICATION/ DIVISION VS. EXPONENTS

Concrete

Solo

1. Have each student complete the investigation individually using the scientific calculator. Make sure to inform the students to input the entire expression into their calculator before hitting the equal key.

Pair Consensus

2. For each problem on the investigation, each student shares with his/her partner, using RallyRobin, his/her response. They discuss the problems that they disagree

trying to come to consensus on the correct response. For the problems they can't reach consensus, they mark those problems so they can focus on them during the team phase. Encourage the students to add to their responses if their partner verbalized an understanding they did not see.

Team Consensus

3. Each pair shares their responses, using RallyRobin, with the other pair in their team, augmenting their responses if necessary. When the teams are through sharing, each student should have a detailed and complete summary.

▶ **Structure**
• Solo-Pair Consensus-Team Consensus

▶ **Materials**
• 1 Blackline 1.3.4 per student
• 1 scientific calculator per student
• 1 pencil per student

5 PROCESSING MULTIPLICATION/ DIVISION VS. EXPONENTS

Numeric

Setup:
In pairs, Student A is the Sage; Student B is the Scribe. Students fold a sheet of paper in half and each writes his/her name on one half.

1. The Sage gives the Scribe step-by-step instructions how to solve problem 1.

2. The Scribe records the Sage's solution step-by-step.

3. If the Sage is correct, the Scribe praises the Sage. Otherwise, the Scribe coaches, then praises.

4. Students switch roles for the next problem.

▶ **Structure**
• Sage-N-Scribe

▶ **Materials**
• 1 Transparency 1.3.5 per pair
• 1 sheet of paper and pencil per pair

WRITE A PROBLEM

▶ **Structure**
• Placemat Consensus

▶ **Materials**
• Transparency 1.3.6
• 1 large sheet of paper (placemat) per team
• 1 pencil per student

Numeric Higher Level Thinking

1. Each team draws a rectangle in the center of the paper that will be the central team space and sections off the area around the large rectangle for individual team member space similar to the diagram below. In the central team space, each team separates it into 3 sections, one for each problem.

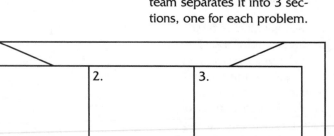

2. Teacher provides an answer and operations (see Transparency 1.3.3).

3. Teammates respond simultaneously in their individual space, writing as many problems as they can using the operations in the order given to get the given answer in the time allotted.

4. Teammate 1 shares with the team one item he/she has written.

5. Teammates discuss the item.

6. If there is consensus that the item is correct, Teammate 1 records his/her problem in the team space of the placemat set aside for problem 1.

7. Repeat steps 4-6, so each teammate in turn suggests an idea and records the team consensus.

8. Repeat steps 2-7 for the remaining problems.

Cooperative Learning and Algebra 1: Becky Bride
Kagan Publishing • 1 (800) 933-2667 • www.KaganOnline.com

7 INVESTIGATING THE ROLE OF GROUPING SYMBOLS

Concrete

Solo

1. Have each student complete the investigation individually using the scientific calculator. Make sure to inform the students to input the entire expression into their calculator before hitting the equal key.

Pair Consensus

2. For each problem on the investigation, each student shares with his/her partner, using RallyRobin, his/her response. They discuss the problems that they disagree trying to come to consensus on the correct response. For the problems they can't reach consensus, they mark those problems so they can focus on them during the team phase. Encourage the students to add to their responses if their partner verbalized an understanding they did not see.

Team Consensus

3. Each pair shares their responses, using RallyRobin, with the other pair in their team, augmenting their responses if necessary. When the teams are through sharing, each student should have a detailed and complete summary.

▶ **Structure**
- Solo-Pair Consensus-Team Consensus

▶ **Materials**
- 1 Blackline 1.3.7 per pair
- 1 scientific calculator per student
- 1 pencil per student

8 PROCESSING ORDER OF OPERATIONS

Numeric

Setup:
In pairs, Student A is the Sage; Student B is the Scribe. Students fold a sheet of paper in half and each writes his/her name on one half.

1. The Sage gives the Scribe step-by-step instructions how to solve problem 1.

2. The Scribe records the Sage's solution step-by-step.

3. If the Sage is correct, the Scribe praises the Sage. Otherwise, the Scribe coaches, then praises.

4. Students switch roles for the next problem.

▶ **Structure**
- Sage-N-Scribe

▶ **Materials**
- Transparency 1.3.8
- 1 sheet of paper and pencil per pair

WHAT DID WE LEARN?

▶ **Structure**
· RoundTable Consensus

▶ **Materials**
· 1 large sheet of paper per team
· different color pen or pencil for each student on the team

Synthesis via Graphic Organizer

1. Each teammate signs his/her name in the upper right corner of the team paper with the color pen/pencil he/she is using.

2. One teammate writes "Order of Operations" in the center of the team paper in a rectangle.

3. Teammate 1 shares with the team one core concept he/she learned in the unit.

4. The student checks for consensus.

5. The teammates show agreement or lack of agreement with thumbs up or down.

6. If there is agreement, the students celebrate and the teammate records the core concept on the graphic organizer connecting it with a line to the main idea, order of operations. If not, teammates discuss the response until there is agreement and then they celebrate.

7. Play continues with the next student's core concept, until all core concepts are exhausted.

8. Repeat steps 3-7 with teammates adding details to each core concept and making bridges between related ideas.

Cooperative Learning and Algebra 1: Becky Bride
Kagan Publishing • 1 (800) 933-2667 • www.KaganOnline.com

ACTIVITY 1 — INVESTIGATING ADDITION/SUBTRACTION VS. MULTIPLICATION/DIVISION

For this investigation, you will need a scientific calculator.

1. a) Enter the problem: $2 + 3 \cdot 4 = ?$ into your calculator and record the answer below.

   b) Without a calculator, add 2 and 3; then multiply the result by 4. Record your answer and work below. Does it match the answer found in part a?

   c) Without a calculator, multiply 3 and 4; then add 2 to the result. Record your answer and work below. Does it match the answer found in part a?

   d) To get the answer in part a, which operation had to be performed first?

2. a) Enter the problem: $5 - 6 \cdot 2 = ?$ into your calculator and record the answer below.

   b) Without a calculator, subtract 6 from 5; then multiply the result by 2. Record your answer and work below. Does it match the answer found in part a?

   c) Without a calculator, multiply 6 and 2; then subtract that result from 5. Record your answer and work below. Does it match the answer found in part a?

   d) To get the answer in part a, which operation had to be performed first?

ACTIVITY 1
INVESTIGATING ADDITION/SUBTRACTION VS. MULTIPLICATION/DIVISION

3. a) Enter the problem: $8 \cdot 5 - 7 = ?$ into your calculator and record the answer below.

   b) Without a calculator, multiply 8 and 5; then subtract 7 from the result. Record your answer and work below. Does it match the answer found in part a?

   c) Without a calculator, subtract 7 from 5; then multiply the result by 8. Record your answer and work below. Does it match the answer found in part a?

   d) To get the answer in part a, which operation had to be performed first?

4. a) Enter the problem: $5 \cdot 7 + 2 = ?$ into your calculator and record the answer below.

   b) Without a calculator, multiply 5 and 7; then add 2 to the result. Record your answer and work below. Does it match the answer found in part a?

   c) Without a calculator, add 7 and 2; then multiply the result by 5. Record your answer and work below. Does it match the answer found in part a?

   d) To get the answer in part a, which operation had to be performed first?

Cooperative Learning and Algebra 1: Becky Bride
Kagan Publishing • 1 (800) 933-2667 • www.KaganOnline.com

ACTIVITY 1 INVESTIGATING ADDITION/SUBTRACTION VS. MULTIPLICATION/DIVISION

5. a) Enter the problem: $6 - 3 \cdot 5 = ?$ into your calculator and record the answer below.

    b) Without a calculator, subtract 3 from 6; then multiply the result by 5. Record your answer and work below. Does it match the answer found in part a?

    c) Without a calculator, multiply 3 and 5; then subtract the result from 6. Record your answer and work below. Does it match the answer found in part a?

    d) To get the answer in part a, which operation had to be performed first?

6. In problems 1-5, which operation was always performed first?

7. a) Enter the problem: $18 \div 2 + 4 = ?$ into your calculator and record the answer below.

    b) Without a calculator, divide 18 by 2; then add 4 to the result. Record your answer and work below. Does it match the answer found in part a?

    c) Without a calculator, add 2 and 4; then divide 18 by the result. Record your answer and work below. Does it match the answer found in part a?

    d) To get the answer in part a, which operation had to be performed first?

INVESTIGATING ADDITION/SUBTRACTION VS. MULTIPLICATION/DIVISION

8. a) Enter the problem: $9 + 12 \div 3 = ?$ into your calculator and record the answer below.

    b) Without a calculator, add 9 and 12; then divide the result by 3. Record your answer and work below. Does it match the answer found in part a?

    c) Without a calculator, divide 12 by 3; then add 9 to the result. Record your answer and work below. Does it match the answer found in part a?

    d) To get the answer in part a, which operation had to be performed first?

9. a) Enter the problem: $28 - 14 \div 7 = ?$ into your calculator and record the answer below.

    b) Without a calculator, subtract 14 from 28; then divide the result by 7. Record your answer and work below. Does it match the answer found in part a?

    c) Without a calculator, divide 14 by 7; then subtract the result from 28. Record your answer and work below. Does it match the answer found in part a?

    d) To get the answer in part a, which operation had to be performed first?

Cooperative Learning and Algebra 1: Becky Bride
Kagan Publishing • 1 (800) 933-2667 • www.KaganOnline.com

INVESTIGATING ADDITION/SUBTRACTION
VS. MULTIPLICATION/DIVISION

10. a) Enter the problem: 36 ÷ 6 + 6 = ? into your calculator and record the answer below.

b) Without a calculator, divide 36 by 6; then add 6 to the result. Record your answer and work below. Does it match the answer found in part a?

c) Without a calculator, add 6 and 6; then divide 36 by the result. Record your answer and work below. Does it match the answer found in part a?

d) To get the answer in part a, which operation had to be performed first?

11. a) Enter the problem: 9 – 15 ÷ 3 = ? into your calculator and record the answer below.

b) Without a calculator, subtract 15 from 9; then divide the result by 3. Record your answer and work below. Does it match the answer found in part a?

c) Without a calculator, divide 15 by 3; then subtract the result from 9. Record your answer and work below. Does it match the answer found in part a?

d) To get the answer in part a, which operation had to be performed first?

12. Summarize what you have learned about the order of operations for addition/subtraction and multiplication/division.

PROCESSING ADDITION/SUBTRACTION VS. MULTIPLICATION/DIVISION

RATIONAL NUMBERS $\frac{-2}{3}$

Structure: Sage-N-Scribe

Simplify each expression.

1. $15 + (-4)(5)$

2. $19 - 24 \div 12$

3. $5 \cdot 3 - 14 \div 7$

4. $6 + (-9) \div 3 - 13$

5. $(-9) - 7 + 3 \cdot 2$

6. $12 \div 4 - 6 \div 2$

7. $5 + 16 \div 4 \cdot (-2)$

8. $9 - (12)(-4) \div 8$

9. $35 \div (-7) - (3)(-5)$

10. $18 \div 2 - (-3)(4)$

Answers:
1. -5
2. 17
3. 13
4. -10
5. -10
6. 0
7. -3
8. 15
9. 10
10. 21

Cooperative Learning and Algebra 1: Becky Bride
Kagan Publishing • 1 (800) 933-2667 • www.KaganOnline.com

3 WRITE A PROBLEM

Structure: Placement Consensus

Write as many problems as you can think of that use the following operations and give the answer requested. The first operation written must also appear first in the problems that you write.

1. Addition, Multiplication, Answer: −5

2. Subtraction, Division, Answer: 12

3. Addition, Division, Multiplication, Answer: −2

INVESTIGATING MULTIPLICATION/ DIVISION VS. EXPONENTS

For this investigation, you will need a scientific calculator.

1. a) Enter the problem: $4 \cdot 3^2 = ?$ into your calculator and record the answer below.
    b) Without a calculator, multiply 4 by 3; then square the result. Record your work and answer below. Does the answer match the answer in part a?

    c) Without a calculator, square 3; then multiply the result by 4. Record your work and answer below. Does the answer match the answer in part a?

    d) To get the answer in part a, which operation had to be performed first?

2. a) Enter the problem: $16 \div 4^2 = ?$ into your calculator and record the answer below.

    b) Without a calculator, divide 16 by 4; then square the result. Record your work and answer below. Does the answer match the answer in part a?

    c) Without a calculator, square 4; then divide 16 by the result. Record your work and answer below. Does the answer match the answer in part a?

    d) To get the answer in part a, which operation had to be performed first?

Cooperative Learning and Algebra 1: Becky Bride
Kagan Publishing • 1 (800) 933-2667 • www.KaganOnline.com

INVESTIGATING MULTIPLICATION/
DIVISION VS. EXPONENTS

3. a) Enter the problem: $8 \div 2^3 = ?$ into your calculator and record the answer below.

    b) Without a calculator, divide 8 by 2; then raise the result to the third power. Record your work and answer below. Does the answer match the answer in part a?

    c) Without a calculator, raise 2 to the third power; then divide 8 by the result. Record your work and answer below. Does the answer match the answer in part a?

    d) To get the answer in part a, which operation had to be performed first?

4. a) Enter the problem: $4 \cdot 3^3 = ?$ into your calculator and record the answer below.

    b) Without a calculator, multiply 4 by 3; then raise the result to the third power. Record your work and answer below. Does the answer match the answer in part a?

    c) Without a calculator, raise 3 to the third power; then multiply the result by 4. Record your work and answer below. Does the answer match the answer in part a?

    d) To get the answer in part a, which operation had to be performed first?

ACTIVITY 4 INVESTIGATING MULTIPLICATION/ DIVISION VS. EXPONENTS

5. a) Enter the problem: $6^2 \div 3 = ?$ into your calculator and record the answer below.

   b) Without a calculator, square 6; then divide the result by 3. Record your work and answer below. Does the answer match the answer in part a?

   c) Without a calculator, divide 6 by 3; then square the result. Record your work and answer below. Does the answer match the answer in part a?

   d) To get the answer in part a, which operation had to be performed first?

6. a) Enter the problem: $8^4 \div 4 = ?$ into your calculator and record the answer below.

   b) Without a calculator, divide 8 by 4; then raise the result to the fourth power. Record your work and answer below. Does the answer match the answer in part a?

   c) Without a calculator, raise 8 to the fourth power; then divide the result by 4. Record your work and answer below. Does the answer match the answer in part a?

   d) To get the answer in part a, which operation had to be performed first?

7. In problems 1-6, what operation was always performed first?

8. Summarize what you have learned about the order of operations for addition/subtraction, multiplication/division, and exponents/roots.

Cooperative Learning and Algebra 1: Becky Bride
Kagan Publishing • 1 (800) 933-2667 • www.KaganOnline.com

ACTIVITY 5
PROCESSING MULTIPLICATION/ DIVISION VS. EXPONENTS

Structure: Sage-N-Scribe

Simplify each expression.

1. $(-6)^2 - \dfrac{-63}{7} + 3^3$

2. $15 + 54 \div 6 - 12$

3. $7 - \dfrac{56}{8} - 3^2$

4. $-5^2 - 3 + 10$

5. $(-9)^2 - 6 \div 2$

6. $-9^2 + 21 \div 3 - 5^2$

7. $(-2)^3 - 8 \times 5 + 3$

8. $-5^2 + \dfrac{20}{5} - (-6)$

Answers:
1. 72
2. 12
3. –9
4. –18
5. 78
6. –99
7. –45
8. –15

ACTIVITY

6 **WRITE A PROBLEM**

Structure: Placement Consensus

Write as many problems as you can think of that use the following operations and give the answer requested.

1. Addition, Exponents, Answer: –5

2. Subtraction, Division, Exponents, Answer: 12

3. Addition, Exponents, Multiplication, Answer: –2

Cooperative Learning and Algebra 1: Becky Bride
Kagan Publishing • 1 (800) 933-2667 • www.KaganOnline.com

INVESTIGATING THE ROLE
OF GROUPING SYMBOLS

For this investigation, you will need a scientific calculator.

1. a) Enter the problem: $5 \cdot (3 + 4) = ?$ into your calculator and record the answer below.

    b) Without a calculator, multiply 5 by 3; then add 4 to the result. Record your work and answer below. Does the answer match the answer in part a?

    c) Without a calculator, add 3 and 4; then multiply the result by 5. Record your work and answer below. Does the answer match the answer in part a?

    d) To get the answer in part a, which operation had to be performed first?

    e) What role did the parentheses play in the order in which the calculator performed the operations?

2. a) Enter the problem: $(5 - 6) \cdot 2 = ?$ into your calculator and record the answer below.

    b) Without a calculator, subtract 6 from 5; then multiply the result by 2. Record your work and answer below. Does the answer match the answer in part a?

    c) Without a calculator, multiply 6 and 2; then subtract the result from 5. Record your work and answer below. Does the answer match the answer in part a?

INVESTIGATING THE ROLE OF GROUPING SYMBOLS

d) To get the answer in part a, which operation had to be performed first?

e) What role did the parentheses play in the order in which the calculator performed the operations?

3. a) Enter the problem: $8 \cdot (5 - 7) = ?$ into your calculator and record the answer below.

b) Without a calculator, multiply 8 by 5; then subtract 7 from the result. Record your work and answer below. Does the answer match the answer in part a?

c) Without a calculator, subtract 7 from 5; then multiply the result by 8. Record your work and answer below. Does the answer match the answer in part a?

d) To get the answer in part a, which operation had to be performed first?

e) What role did the parentheses play in the order in which the calculator performed the operations?

4. a) Enter the problem: $(5 + 2)^2 = ?$ into your calculator and record the answer below.

b) Without a calculator, add 5 and 2; then square the result. Record your work and answer below. Does the answer match the answer in part a?

c) Without a calculator, square 2; then add 5 to the result. Record your work and answer below. Does the answer match the answer in part a?

Cooperative Learning and Algebra 1: Becky Bride
Kagan Publishing • 1 (800) 933-2667 • www.KaganOnline.com

INVESTIGATING THE ROLE OF GROUPING SYMBOLS

d) To get the answer in part a, which operation had to be performed first?

e) What role did the parentheses play in the order in which the calculator performed the operations?

5. a) Enter the problem: $18 \div (2 + 4) = ?$ into your calculator and record the answer below.

    b) Without a calculator, divide 18 by 2; then add 4 to the result. Record your work and answer below. Does the answer match the answer in part a?

    c) Without a calculator, add 2 and 4; then divide 18 by the result. Record your work and answer below. Does the answer match the answer in part a?

    d) To get the answer in part a, which operation had to be performed first?

    e) What role did the parentheses play in the order in which the calculator performed the operations?

6. a) Enter the problem: $(9 + 12) \div 3 = ?$ into your calculator and record the answer below.

    b) Without a calculator, add 9 and 12; then divide the result by 3. Record your work and answer below. Does the answer match the answer in part a?

INVESTIGATING THE ROLE
OF GROUPING SYMBOLS

c) Without a calculator, divide 12 by 3; then add 9 to the result. Record your work and answer below. Does the answer match the answer in part a?

d) To get the answer in part a, which operation had to be performed first?

e) What role did the parentheses play in the order in which the calculator performed the operations?

7. a) Enter the problem: $(28 - 14) \div 7 = ?$ into your calculator and record the answer below.

b) Without a calculator, subtract 14 from 28; then divide the result by 7. Record your work and answer below. Does the answer match the answer in part a?

c) Without a calculator, divide 14 by 7; then subtract the result from 28. Record your work and answer below. Does the answer match the answer in part a?

d) To get the answer in part a, which operation had to be performed first?

e) What role did the parentheses play in the order in which the calculator performed the operations?

8. a) Enter the problem: $\dfrac{9}{(2-5)} = ?$ into your calculator and record the answer below.

b) To get the answer above, which operation had to be performed first?

Cooperative Learning and Algebra 1: Becky Bride
Kagan Publishing • 1 (800) 933-2667 • www.KaganOnline.com

INVESTIGATING THE ROLE OF GROUPING SYMBOLS

c) What role did the parentheses play in the order in which the calculator performed the operations?

9. a) Enter the problem: $\dfrac{(9-15)}{3}$ = ? into your calculator and record the answer below.

b) To get the answer above, which operation had to be performed first?

c) What role did the parentheses play in the order in which the calculator performed the operations?

10. a) Enter the problem: $\dfrac{24}{(1+2)}$ = ? into your calculator and record the answer below.

b) To get the answer above, which operation had to be performed first?

c) What role did the parentheses play in the order in which the calculator performed the operations?

11. Summarize what you have learned about the role of grouping symbols in the order operations on numbers are performed.

12. Based on all the investigations you have done on order of operations, make a list of the order operations are performed in.

ACTIVITY 8 PROCESSING ORDER OF OPERATIONS

Structure: Sage-N-Scribe

Simplify each expression.

1. $5(3+6)-4$

2. $4 - \dfrac{2^3 - 3}{5+1} + (-10)$

3. $(-20) + \dfrac{3(2-8)}{|-9|}$

4. $3(5 \div 2 - 1) + |6|$

5. $(3^2 - 4)(4 - 3^2)$

6. $\dfrac{(5 + 4^2)(2 - 3^2)}{(4) + 6}$

7. $|4 - 7| + 7^2 \div 2$

8. $8^2 - |3 - 24 \div 6|$

Answers:
1. 41
2. $\dfrac{-41}{6}$
3. -22
4. $\dfrac{21}{2}$
5. -25
6. $\dfrac{-147}{10}$
7. $\dfrac{55}{2}$
8. 63

Cooperative Learning and Algebra 1: Becky Bride
Kagan Publishing • 1 (800) 933-2667 • www.KaganOnline.com

EXPRESSIONS

This chapter begins with a lesson on vocabulary. The first activity requires students to generate definitions. Several activities follow to process the definitions. One goal is for students to understand the difference between a "coefficient" and an "exponent" and what each of them means. Students will write expressions using words and then symbols. Students will also see how expressions are used in real-world applications.

The second lesson works with evaluating expressions. One activity ties geometry to algebra by requiring students to evaluate geometric expressions. A higher-level thinking activity requires the students to find the mistake in an expression that has been evaluated. This lesson concludes with real-world applications of expressions.

The third lesson focuses on simplifying expressions. Two exploratory activities have students discover like and unlike terms and the distributive property. Several activities are included to process each of the investigations. This lesson ends with an activity where the students develop a graphic organizer that summarizes the entire lesson.

The fourth lesson focuses on exponents and the rules of exponents. Explorations have students discover the product, quotient, power rules of exponents, and negative and zero exponents. Activities are included to process each of these explorations. The final activities revisit common factoring and the distributive property. The chapter concludes with an activity to synthesize the unit.

LESSON 1 VOCABULARY DEVELOPMENT
ACTIVITY 1: Define Me!
ACTIVITY 2: Name My Coefficient
ACTIVITY 3: Identify My Terms
ACTIVITY 4: Name the Factors
ACTIVITY 5: Can You Write Me?
ACTIVITY 6: Expression, Equation, or Inequality? Oh My!
ACTIVITY 7: Write Me Using Words
ACTIVITY 8: Write Me as an Expression
ACTIVITY 9: Apply Expressions
ACTIVITY 10: What Did We Learn?

LESSON 2 EVALUATING EXPRESSIONS
ACTIVITY 1: Evaluate My Expression
ACTIVITY 2: Evaluate My Geometric Expression
ACTIVITY 3: Where Are the Mistakes?
ACTIVITY 4: Expressions in the Real World

EXPRESSIONS

LESSON 3

SIMPLIFYING EXPRESSIONS

ACTIVITY 1: Can You Form Me with Tiles?
ACTIVITY 2: Can You Put Me Together?
ACTIVITY 3: Can You Write My Expression?
ACTIVITY 4: Exploring Like and Unlike Terms
ACTIVITY 5: Find My Like Terms
ACTIVITY 6: Simplify Me With Tiles
ACTIVITY 7: Simplify Me Without Tiles
ACTIVITY 8: Make Me with Tiles
ACTIVITY 9: Make Me with Tiles—Advanced
ACTIVITY 10: Exploring the Distributive Property
ACTIVITY 11: Simplify Me with the Distributive Property
ACTIVITY 12: Find My Opposite
ACTIVITY 13: Simplify Me—Advanced
ACTIVITY 14: Factor Me—Distributive Property Undone
ACTIVITY 15: Create an Application
ACTIVITY 16: What Did We Learn?

LESSON 4

EXPONENTS

ACTIVITY 1: Expand Me
ACTIVITY 2: Rewrite Me
ACTIVITY 3: Exploring the Product Rule of Exponents
ACTIVITY 4: Use the Product Rule
ACTIVITY 5: Exploring the Quotient Rule of Exponents
ACTIVITY 6: Use the Quotient Rule
ACTIVITY 7: Exploring the Power of a Product/Quotient Rule
ACTIVITY 8: Use the Power of a Product/Quotient Rule
ACTIVITY 9: Exploring Negative and Zero Exponents
ACTIVITY 10: Rewrite Me with Positive Exponents
ACTIVITY 11: Distributive Property Revisited
ACTIVITY 12: Factor Me—Revisited
ACTIVITY 13: What Did We Learn?

Cooperative Learning and Algebra 1: Becky Bride
Kagan Publishing • 1 (800) 933-2667 • www.KaganOnline.com

LESSON 1
VOCABULARY DEVELOPMENT

This lesson begins with an activity that requires students to generate definitions. Because the words "coefficient," "factors," and "terms" are vocabulary words students often confuse, student-generated definitions become even more important. The activity is designed so students not only can identify these items in a problem but that they also understand what these terms mean. Also included are the words "sum," "difference," "product," and "quotient." In the expressions that are examples of these terms, parentheses have been used because when translating from words to symbols parentheses are needed. Several activities follow to process the definitions. The activity, "Can You Write Me?" has the students apply the definitions as they create terms, expressions, equations, or inequalities, based on given criteria. Near the end of the lesson, students will take expressions written with words and write them symbolically. Then students will take a symbolic form of an expression and write it using words. The lesson ends with an activity to synthesize what they have learned by developing a graphic organizer.

ACTIVITY

1 DEFINE ME!

Exploratory

Solo

1. Individually, each student defines each vocabulary word by comparing and contrasting the examples and counterexamples. Students could do this part for homework.

Team with RoundRobin sharing and RoundTable recording.

2. In turn, each student reads to his/her team the definition he/she wrote.

3. The team discusses how to define the vocabulary word, based on the definitions just shared. The team must come to consensus on how to define the term.

4. Once the team reaches consensus, student 1 records the team definition on the team paper.

5. Repeat steps 2-4 for each vocabulary word, rotating the recording.

Class

6. Choose a team and a student at that team to read their team's definition of the first vocabulary word. (Team and student spinners work well here.)

7. Write the definition on the board. Ask the class if they agree with part 1, then part 2, and finally part 3 of the definition, reworking what they want to change.

▶ **Structure**
• Solo Team Class

▶ **Materials**
• 1 Blackline 2.1.1 per student
• 1 sheet of paper and pencil per student

8. Once everyone has agreed, including the teacher, then it is recorded as the definition the class will use. You may find that lower-level classes prefer a somewhat longer definition if it has more meaning for them. Honors classes will try to be as concise as possible.

ACTIVITY 2 NAME MY COEFFICIENT

▶ **Structure**
• RallyRobin

▶ **Materials**
• Transparency 2.1.2

Algebraic

1. Teacher puts transparency 2.1.2 on the overhead projector that presents several expressions.

2. In pairs, students take turns orally stating the coefficient of the term.

ACTIVITY 3 IDENTIFY MY TERMS

▶ **Structure**
• RallyRobin

▶ **Materials**
• Transparency 2.1.3

Algebraic

1. Using Transparency 2.1.3, several expressions are written.

2. In pairs, students take turns orally stating each term of the expression.

ACTIVITY 4 NAME THE FACTORS

▶ **Structure**
• RallyRobin

▶ **Materials**
• Transparency 2.1.3

Algebraic

1. Using transparency 2.1.3, several expressions are written.

2. In pairs, students take turns orally stating the factors in each term of the expression.

Cooperative Learning and Algebra 1: Becky Bride
Kagan Publishing • 1 (800) 933-2667 • www.KaganOnline.com

ACTIVITY

5 CAN YOU WRITE ME?

Algebraic Higher-Level Thinking

Setup:
In pairs, Student A is the Sage; Student B is the Scribe. Students fold a sheet of paper in half and each writes his/her name on one half.

1. The Sage gives the Scribe step-by-step instructions on how to write the expression for problem 1.

2. The Scribe records the Sage's expression step-by-step in writing.

3. If the Sage is correct, the Scribe praises the Sage. Otherwise, the Scribe coaches, then praises.

4. Students switch roles for the next problem.

▶ **Structure**
• Sage-N-Scribe

▶ **Materials**
• 1 Blackline 2.1.4 per pair
• 1 sheet of paper and pencil per pair

ACTIVITY

6 EXPRESSION, EQUATION, OR INEQUALITY? OH MY!

Algebraic

Setup:
The cards need to be copied onto cardstock and cut into individual cards. Cards are distributed to the students, 1 card per student.

1. Stand Up-Hand Up-Pair Up

2. Partner A quizzes his/her partner, asking what type of mathematical sentence is on his/her card.

3. Partner B answers.

4. Partner A praises or coaches.

5. Switch roles.

6. Partners trade cards.

7. Repeat steps 1-6.

Management Tip: Using music to begin and stop the mixing works well.

▶ **Structure**
• Quiz-Quiz-Trade

▶ **Materials**
• 1 set of expression, equation, inequality cards (Blackline 2.1.5)

Chapter 2: Expressions
Lesson One

ACTIVITY

7 WRITE ME USING WORDS

▶ **Structure**
• Sage-N-Scribe

▶ **Materials**
• Transparency 2.1.6
• 1 sheet of paper and pencil per pair

Algebraic

Setup:
In pairs, Student A is the Sage; Student B is the Scribe. Students fold a sheet of paper in half and each writes his/her name on one half.

1. The Sage gives the Scribe step-by-step instructions on how to write the expression in words.

2. The Scribe records the Sage's expression word-for-word in writing.

3. If the Sage is correct, the Scribe praises the Sage. Otherwise, the Scribe coaches, then praises.

4. Students switch roles for the next problem.

ACTIVITY

8 WRITE ME AS AN EXPRESSION

▶ **Structure**
• Sage-N-Scribe

▶ **Materials**
• Transparency 2.1.7
• 1 sheet of paper and pencil per pair

Algebraic

Setup:
In pairs, Student A is the Sage; Student B is the Scribe. Students fold a sheet of paper in half and each writes his/her name on one half.

1. The Sage gives the Scribe step-by-step instructions on how to write the expression using variables and numbers.

2. The Scribe records the Sage's expression.

3. If the Sage is correct, the Scribe praises the Sage. Otherwise, the Scribe coaches, then praises.

4. Students switch roles for the next problem.

Cooperative Learning and Algebra 1: Becky Bride
Kagan Publishing • 1 (800) 933-2667 • www.KaganOnline.com

ACTIVITY
9 APPLY EXPRESSIONS

Connect

Setup:
In pairs, Student A is the Sage; Student B is the Scribe. Students fold a sheet of paper in half and each writes his/her name on one half.

1. The Sage gives the Scribe step-by-step instructions on how to solve problem 1.

2. The Scribe records the Sage's solution step-by-step.

3. If the Sage is correct, the Scribe praises the Sage. Otherwise, the Scribe coaches, then praises.

4. Students switch roles for the next problem.

Answers to Activity 9:
1. x = number of months; 150 + 475x
2. x = number of months; 199 + 688(x - 1)
3. n = number of people after the first person; 20 + 15n
4. n = number of games played; 10 + .25n

▶ **Structure**
· Sage-N-Scribe

▶ **Materials**
· 1 Blackline 2.1.8 per pair
· 1 sheet of paper and pencil per pair

ACTIVITY
10 WHAT DID WE LEARN?

Synthesis via Graphic Organizer

1. Each teammate signs his/her name in the upper-right corner of the team paper with the color pen/pencil he/she is using.

2. One teammate writes "Expression Vocabulary" in the center of the team paper in a rectangle.

3. Teammate 1 shares with the team one core concept he/she learned in the unit.

4. The student checks for consensus.

5. The teammates show agreement or lack of agreement with thumbs up or down.

6. If there is agreement, the students celebrate and the teammate records the core concept on the graphic organizer, connecting it with a line to the main idea, "Expression Vocabulary." If not, teammates discuss the response until there is agreement and then they celebrate.

7. Play continues with the next student's core concept, until all core concepts are exhausted.

8. Repeat steps 3-7, with teammates adding details to each core concept and making bridges between related ideas.

▶ **Structure**
· RoundTable Consensus

▶ **Materials**
· 1 large sheet of paper per team
· different color pen or pencil for each student on the team

ACTIVITY

1 DEFINE ME!

For each vocabulary word below, compare and contrast the examples with the counterexamples. Write a definition based on the similarities and differences you notice.

1. Coefficient

Examples	Counterexamples
2x; 2 is a coefficient and it means x + x	2x; x is not a coefficient
$4x^8$; 4 is a coefficient and it means $x^8 + x^8 + x^8 + x^8$	$4x^8$; x and 8 are not coefficients
9xy; 9 is a coefficient and it means xy + xy + xy + xy + xy + xy + xy + xy + xy	9xy; x and y are not coefficients
3x; 3 is a coefficient and it means x + x + x	3x; x is not a coefficient

2. Term

Examples	Counterexamples
5x + 3; 5x and 3 are terms	5x + 3; 5 is not a term and x is not a term
$6x^2y$ – 4x; $6x^2y$ and 4x are terms	$6x^2y$ + 4x; 6 is not a term, 6y is not a term, 4 is not a term, $6x^2$ is not a term, x^2y is not a term
$5x^2$ – 3x + 7; $5x^2$ and 3x and 7 are terms	$5x^2$ – 3x + 7; 5, 3, and 7 are not terms, x^2 is not a term, x is not a term

Cooperative Learning and Algebra 1: Becky Bride
Kagan Publishing • 1 (800) 933-2667 • www.KaganOnline.com

ACTIVITY

1 DEFINE ME!

3. Factors

Examples	Counterexamples
$5xy$; 5, x and y are factors	$5 + x + y$; 5, x and y are not factors
$7x^3y^2w$; 7, x^3, y^2, and w are factors	$7 - x^3 + y^2 - w$; 7, x^3, y^2 and w are not factors
$3x + 6y$; 3 and x are factors; 6 and y are factors	$3x + 6y$; 3x and 6y are not factors

4. Expressions

Examples	Counterexamples
$5x + y^2$	$5x + y^2 = 10$
$6xy^3z$	$6xy^3z \leq 20$
$5(6x + 3) - 4x^2$	$5(6x + 3) - 4x^2 > 8$
$\dfrac{x + 7y}{p}$	$\dfrac{x + 7y}{p} = 15$

5. Equations

Examples	Counterexamples
$5x + y^2 = 10$	$5x + y^2$
$\dfrac{x + 7y}{p} = 15$	$6xy^3z$
$5x + 3 = 14 - 6x$	$6xy^3z \leq 20$
$6x^2 - 5x + 3 = 12$	$5(6x + 3) - 4x^2$

DEFINE ME!

6. Inequalities

Examples	Counterexamples
$6xy^3z \leq 20$	$6x^2 - 5x + 3 = 12$
$5(6x +3) - 4x^2 > 8$	$5(6x +3) - 4x^2$
$6x \geq (-3)$	$\dfrac{x+7y}{p} = 15$
$27x - 72 < 5x + 3$	$5x + y^2$

7. Sum

Examples	Counterexamples
The sum of 2 and 3 is (2 + 3)	(2 - 3) is not the sum of 2 and 3
The sum of 4 and -5 is [4 + (-5)]	[4 ÷ (-5)] is not the sum of 4 and -5
The sum of x and 3 is (x + 3)	3x is not the sum of x and 3

8. Difference

Examples	Counterexamples
The difference of 4 and 1 is (4 − 1)	(4 + 1) is not the difference of 4 and 1
The difference of r and 2 is (r − 2)	2r is not the difference of r and 2
The difference of −8 and 7 is (−8 − 7)	(−8 + 7) is not the difference of −8 and 7

Cooperative Learning and Algebra 1: Becky Bride
Kagan Publishing • 1 (800) 933-2667 • www.KaganOnline.com

ACTIVITY

1 **DEFINE ME!**

9. Product

Examples	**Counterexamples**
The product of 4 and 3 is (4)(3)	(4 + 3) is not the product of 4 and 3
The product of 3 and w is 3w	$\frac{3}{w}$ is not the product of 3 and w
The product of −1 and 5 is (−1)(5)	(−1 − 5) is not the product of −1 and 5

10. Quotient

Examples	**Counterexamples**
The quotient of 3 and 4 is $\frac{3}{4}$	(3 + 4) is not the quotient of 3 and 4
The quotient of 8 and n is $\frac{8}{n}$	(8 − n) is not the quotient of 8 and n
The quotient of −5 and 2 is $\frac{-5}{2}$	(−5)(2) is not the quotient of −5 and 2

2 **NAME MY COEFFICIENT**

Structure: RallyRobin

Name the coefficient and explain what it means.

1. $3x^2$

2. $5h^6$

3. $6p^3$

4. $4x^2y^5$

5. $3x^2m$

6. $8g^4xy$

7. $-5k^2wq^3$

8. $-7r^8ph$

Answers:
1. 3; $x^2 + x^2 + x^2$
2. 5; $h^6 + h^6 + h^6 + h^6 + h^6$
3. 6; $p^3 + p^3 + p^3 + p^3 + p^3 + p^3$
4. 4; $x^2y^5 + x^2y^5 + x^2y^5 + x^2y^5$
5. 3; $x^2m + x^2m + x^2m$
6. 8; $g^4xy + g^4xy + g^4xy + g^4xy + g^4xy + g^4xy + g^4xy + g^4xy$
7. -5; $-k^2wq^3 - k^2wq^3 - k^2wq^3 - k^2wq^3 - k^2wq^3$
8. -7; $-r^8ph - r^8ph - r^8ph - r^8ph - r^8ph - r^8ph - r^8ph$

Cooperative Learning and Algebra 1: Becky Bride
Kagan Publishing • 1 (800) 933-2667 • www.KaganOnline.com

ACTIVITY 3

IDENTIFY MY TERMS

Structure: RallyRobin

Activity 3: Identify each term in the expression.
Activity 4: Name the factors in each term.

1. $3x^2 - 4m + 2$

2. $7x^2r^4p^3 + 3p$

3. $8m^2hw^6$

4. $8g + xy - 4p$

5. $3zg^4 + xyw^2 + 4p$

6. $-5k^2wq$

7. $-7m^2y + 3d$

8. $4mht^2 + 5y - 7x$

Activity 3 Answers:
1. $3x^2$; $-4m$; 2
2. $7x^2r^4p^3$; $3p$
3. $8m^2hw^6$
4. $8g$; xy; $-4p$
5. $3zg^4$; xyw^2; $4p$
6. $-5k^2wq$
7. $-7m^2y$; $3d$
8. $4mht^2$; $5y$; $-7x$

Activity 4 Answers:
1. 3 and x^2; 4 and m
2. 7, x^2, r^4, and p^3; 3 and p
3. 8, m^2, h, and w^6
4. 8 and g^4, x and y, 4 and p
5. 3, z, g^4; x, y, and w^2; 4 and p
6. -5, k^2, w and q
7. -7, m^2, and y; 3 and d
8. 4, m, h, and t^2; 5 and y; 7 and x

5 CAN YOU WRITE ME?

Structure: Sage-N-Scribe

Write an expression that meets the criteria in each problem.

1. I have one term. The coefficient is five and I have two other factors, one of which is squared.

2. I have two terms. The coefficient of the first term is eight and it has one additional factor that is raised to the fifth power. The second term has a coefficient of negative seven and two other factors, one squared and the other cubed.

3. I have three terms. The coefficient of the first term is five and it has two other factors one of which is raised to the fourth power. The second term has a coefficient of negative three and another factor that is raised to the sixth power. The coefficient of the third term is one and it has no factors.

4. I have two terms. The coefficient of the first term is a fraction whose value is between three and four and has one additional factor that is squared. The coefficient of the second term is a decimal between negative two and negative three and it has three additional factors. One factor is raised to the seventh power, another factor is raised to the eighth power, and the other factor is squared.

Sample Answers:
1. $5xy^2$
2. $8x^5 - 7x^2y^3$
3. $5ab^4 - 3a^6 + 1$
4. $\dfrac{7}{2}a^2 - 2.5a^7b^8c^2$

Cooperative Learning and Algebra 1: Becky Bride
Kagan Publishing • 1 (800) 933-2667 • www.KaganOnline.com

ACTIVITY 6
EXPRESSION, EQUATION, OR INEQUALITY? OH MY!

EXPRESSIONS $2m - 4$

Structure: Quiz-Quiz-Trade

$5x + 3$	$4x - 5 = 3$	$9x \leq 5$
$x^2 - 3$	$3x(x + 2) = 1$	$5x - 9 \geq 3$
$x + 7y$	$2x - 5 = 3x$	$y < 5$
$6(y + 2x)$	$x - 1 = x + 6$	$x + 6 < 5 - x$
$ab^3 + 5b$	$m^2 - 5m + 3 = 4$	$7h - 4 > 9$
$2x + 3y$	$4x - 15 = 3 + x$	$9x - 8 \leq 5$

ACTIVITY 6
EXPRESSION, EQUATION, OR INEQUALITY? OH MY!

EXPRESSIONS
$2m-4$

Structure: Quiz-Quiz-Trade

$5x^2 - 13x$	$3x(4x + 2) = 1y$	$5x + w \leq 8$
$5x^2 - 13x$	$2x^3 - 9 = 3x2$	$y - x < 5$
$6(y + 2h) + 3h$	$5x - 1 = x^2$	$3x + 7 < 5 - 2x$
$5a(ab^3 + 5b)$	$m^2 - 2m + 9 = 4$	$7h^2 - 4 > 9h$
$x^2 + 7pw - 4m$	$7w = 3$	$20 \leq g$
$8x^2 + 7p - 9xp$	$7w - 4f = 3c$	$20 \leq g \leq 50$

Cooperative Learning and Algebra 1: Becky Bride
Kagan Publishing • 1 (800) 933-2667 • www.KaganOnline.com

7 WRITE ME USING WORDS

Structure: Sage-N-Scribe

Write each expression using words.

1. $x + 3$

2. $-6k$

3. $\dfrac{m}{9}$

4. $8 - e$

5. $2m - 4$

6. $5(3h + 8)$

7. $\dfrac{4p - 5}{7}$

8. $2 + \dfrac{3}{d}$

Answers:
1. Some number increased by three.
2. Some number multiplied by negative six.
3. Some number divided by nine.
4. Some number less than eight.
5. Twice some number decreased by four.
6. Five times the sum of three times some number and eight.
7. The difference of four times some number and five divided by seven.
8. Two increased by three divided by some number.

8 WRITE ME AS AN EXPRESSION

EXPRESSIONS
$2m-4$

Structure: Sage-N-Scribe

a) Define a variable.
b) Write each of the following as an expression.

1. A number increased by seven

2. The sum of two and a number divided by three

3. The quotient of nine and some number decreased by six

4. Five less than twice a number

5. The product of a number and eight divided by five

6. The difference of a number and two multiplied by seven

Answers:

1. $n + 7$

2. $\dfrac{2+n}{3}$

3. $\dfrac{9}{n} - 6$

4. $2n - 5$

5. $\dfrac{8n}{5}$

6. $7(n - 2)$

Cooperative Learning and Algebra 1: Becky Bride
Kagan Publishing • 1 (800) 933-2667 • www.KaganOnline.com

9 APPLY EXPRESSIONS

EXPRESSIONS
2m-4

Structure: Sage-N-Scribe

For each problem, a) define a variable and b) write an expression that models the problem.

1. Write an expression for the total amount of money spent on housing for a 3-bedroom apartment for numerous months.

	MOVE IN NOW!!	
	OVER 600 HOMES	
4br/1ba	$600 dep.	$700 mth
3br/1ba	$150 dep.	$475 mth
2br/1ba	$100 dep.	$360 mth
1br/1ba	$100 dep.	$320 mth

2. Write an expression for the total amount of money spent on housing for a 2-bedroom apartment for numerous months.

WOW!!!
$199 Move In!
2 bedroom 1.5 bath
1,100 square feet, peaceful setting in a quiet pet friendly community. $688 mth
1 Month FREE!

3. Write an expression for the total amount of money spent for an airport shuttle ride for numerous people.

BEST RATES IN TOWN!!!
Airport shuttle—first person $20 and each additional person $15.

4. Write an expression for the total amount of money spent on a day at an amusement park for numerous arcade games.

JULY SPECIAL!
Get out of the heat and play your favorite games!
Admission to the park $10 and each arcade game is only $0.25.

Answers on teacher notes (page 83).

LESSON 2
EVALUATING EXPRESSIONS

This lesson processes the evaluating of algebraic expressions. Included is an activity that connects algebra with geometry where students evaluate geometric expressions involving area, perimeter, and volume. A higher-level thinking activity is included, which requires students to find the mistake(s) of an expression that was evaluated. This lesson ends with applications of expressions in a real-world context. These applications reinforce the concept of "variable" and writing expressions.

ACTIVITY 1
EVALUATE MY EXPRESSION

▶ **Structure**
• Sage-N-Scribe

▶ **Materials**
• Transparency 2.2.1
• 1 sheet of paper and pencil per pair

Algebraic

Setup:
In pairs, Student A is the Sage; Student B is the Scribe. Students fold a sheet of paper in half and each writes his/her name on one half.

1. The Sage gives the Scribe step-by-step instructions how to solve problem 1.

2. The Scribe records the Sage's solution step-by-step in writing.

3. If the Sage is correct, the Scribe praises the Sage. Otherwise, the Scribe coaches, then praises.

4. Students switch roles for the next problem.

ACTIVITY 2
EVALUATE MY GEOMETRIC EXPRESSION

▶ **Structure**
• Sage-N-Scribe

▶ **Materials**
• 1 Blackline 2.2.2 per pair
• 1 sheet of paper and pencil per pair

Connect Algebra to Geometry

Setup:
In pairs, Student A is the Sage; Student B is the Scribe. Students fold a sheet of paper in half and each writes his/her name on one half.

1. The Sage gives the Scribe step-by-step instructions on how to solve problem 1.

2. The Scribe records the Sage's solution step-by-step in writing.

3. If the Sage is correct, the Scribe praises the Sage. Otherwise, the Scribe coaches, then praises.

4. Students switch roles for the next problem.

ACTIVITY

3 WHERE ARE THE MISTAKES?

Algebraic Higher-Level Thinking

1. Each team draws a rectangle in the center of the paper, which will be the central team space, and sections off the area around the large rectangle for individual team member space similar to the diagram below. In the central team space, each team separates it into 3 sections, one for each problem.

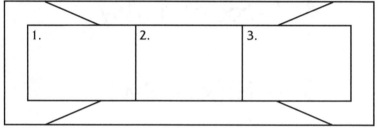

2. Teacher puts Transparency 2.2.3 on the overhead projector.

3. Teammates all respond simultaneously in their individual space, finding and recording as many mistakes as they can in the allotted time.

4. Teammate 1 shares with the team one item he/she has written.

5. Teammates discuss the item.

6. If there is consensus that the item is correct, teammate 1 records his/her application in the team space of the placemat set aside for problem 1.

7. Repeat steps 4-6, so each teammate, in turn, suggests an idea and records the team consensus.

8. Repeat steps 2-7 for the remaining problems, changing transparencies when necessary.

▶ **Structure**
• RoundTable Consensus

▶ **Materials**
• Transparencies 2.2.3, 2.2.4, and 2.2.5
• 1 large sheet of paper
• 1 pen/pencil per student

ACTIVITY

4 EXPRESSIONS IN THE REAL WORLD

Connect

Setup:
In pairs, Student A is the Sage; Student B is the Scribe. Students fold a sheet of paper in half and each writes his/her name on one half.

1. The Sage gives the Scribe step-by-step instructions on how to do problem 1.

2. The Scribe records the Sage's solution step-by-step in writing.

3. If the Sage is correct, the Scribe praises the Sage. Otherwise, the Scribe coaches, then praises.

4. Students switch roles for the next problem.

Sample answers to Activity 4:
1. $f + 6n$; setup fee \$40 and number of t-shirts 30; \$220

2. $w + b$; $w = \$32000$ and $b = \$500$; \$32,500

3. hp; $h = 40$ and $p = \$6$ per hour; \$240

4. $40p + 21c$; $p = \$6$ per hour and $c = \$10$; \$450

▶ **Structure**
• Sage-N-Scribe

▶ **Materials**
• Transparency 2.2.6
• 1 sheet of paper and pencil per pair

5. $\dfrac{f}{4000}$; $f = 40,000$; 10 bags

6. $12p$; \$700; \$8,400

7. $b + d - c$; $b = \$200$, $d = \$340$, and $c = \$280$; \$260

8. $1.07p$; $p = \$30$; \$32.10

9. $0.85p - 200$; $p = \$1,000$; \$650

10. $12u + 8n$; $u = \$7$ and $n = \$18$; \$228

ACTIVITY 1 — EVALUATE MY EXPRESSION

Structure: Sage-N-Scribe

Evaluate each expression if $x = 2$, $y = (-3)$, $w = (-4)$, $p = \dfrac{1}{2}$, and $m = 1.2$.

1. $2x + 3y - 4w$

2. $x^2 - y^2$

3. $(2p)^3 - \dfrac{w}{x}$

4. $(-5xm) + 4w$

5. $5(3p + y)$

6. $(-8)(x - 2w)$

7. $\dfrac{2w^2}{x^2 + 3y}$

8. $(5 + [2x - w]) - 2$

9. $(w + x)(w - x)$

10. $\dfrac{5w - y}{x^2}$

Answers:
1. 11
2. −5
3. 3
4. −28
5. −7.5
6. −80
7. $\dfrac{-32}{5}$
8. 11
9. 12
10. $\dfrac{-17}{4}$

Cooperative Learning and Algebra 1: Becky Bride
Kagan Publishing • 1 (800) 933-2667 • www.KaganOnline.com

EVALUATE MY GEOMETRIC EXPRESSION

EXPRESSIONS
2m-4

Structure: Sage-N-Scribe

Evaluate each expression. Round to the hundredths place.

1. The perimeter of a rectangle is 2l + 2w, where l is the length of the rectangle and w is the width. Find the perimeter of the rectangle if the length is 12 inches and the width is 9 inches.

2. The area of a triangle is $\frac{1}{2}$bh, where b is the base of the triangle and h is the height of the triangle. Find the area of a triangle whose base is 13 m and height is 8 m.

3. The surface area of a sphere is $4\pi r^2$, where π is approximately 3.14 and r is the radius. Find the surface area of a spherical balloon whose radius is 7 cm.

4. The sum of the measures of the interior angles of a polygon is 180(n − 2), where n is the number of sides of the polygon. Find the sum of the measures of the interior angles of a polygon that has 9 sides.

5. The volume of a cone is $\frac{1}{3}\pi r^2 h$, where π is approximately 3.14, r is the radius, and h is the height. Find the volume of an ice-cream cone if its radius is 1.5 inches and its height is 6 inches.

EVALUATE MY GEOMETRIC EXPRESSION

EXPRESSIONS
2m-4

Structure: Sage-N-Scribe

6. The area of a trapezoid is $\frac{1}{2}h(b_1+b_2)$, where h is the height, b_1 is one base, and b_2 is the other base. Find the area of the trapezoid whose height is 3 cm, one base measures 8 cm, and the other base measures 15 cm.

7. The surface area of a rectangular prism is 2lw + 2lh + 2wh, where l is the length, w is the width, and h is the height. Find the surface area of a prism whose length is 5 feet, width is 8 feet, and height is 9 feet.

8. The volume of a sphere is $\frac{4}{3}\pi r^3$, where π is approximately 3.14 and r is the radius. Find the volume of a ball whose radius is 12 cm.

Answers:
1. 42 in.
2. 52 m²
3. 615.44 cm²
4. 1260°
5. 14.13 in³
6. 34.5 cm²
7. 314 ft²
8. 7234.56 cm³

Cooperative Learning and Algebra 1: Becky Bride
Kagan Publishing • 1 (800) 933-2667 • www.KaganOnline.com

WHERE ARE THE MISTAKES?

Structure: RoundTable Consensus

Use the expression: $5(3h + m) - 4d^2$, where $h = 2$, $m = 4$, and $d = (-1)$.

$$5(3h + m) - 4d^2 = 5[3(2) + 4] - 4 \cdot -1^2$$
$$5[6 + 4] + 4^2$$
$$5[10] + 8$$
$$50 + 8$$
$$58$$

ACTIVITY
3 **WHERE ARE THE MISTAKES?**

Structure: RoundTable Consensus

Use the expression: $48 \div y \cdot x - 4k$, where $y = -2$, $x = 4$, and $k = (-3)$.

$48 \div y \cdot x - 4k = 48 \div (-2) \cdot 4 - 4(-3)$
$ 48 \div (-8) - 4\,(-3)$
$ -6 - 4\,(-3)$
$ -10\,(-3)$
$ 30$

Cooperative Learning and Algebra 1: Becky Bride
Kagan Publishing • 1 (800) 933-2667 • www.KaganOnline.com

WHERE ARE THE MISTAKES?

EXPRESSIONS
2m-4

Structure: RoundTable Consensus

Use the expression: $\dfrac{2[w-(-h)]}{p}+5p^2$,

where w = –3, h = –5, and p = 2.

$$\dfrac{2[w-(-h)]}{p}+5p^2 = \dfrac{2[-3-(-5)]}{2}+5(2)^2$$

$$= \dfrac{2[-8]}{2}+5(2)^2$$

$$= \dfrac{-16}{2}+5(2)^2$$

$$= -8+10^2$$

$$= 2^2$$

$$= 4$$

ACTIVITY
4 EXPRESSIONS IN THE REAL WORLD

Structure: Sage-N-Scribe

For each problem:
   a) Write an expression to model the problem.
   b) Give a reasonable value for each variable.
   c) Evaluate your expression for the value(s) you chose in part b.

1. f = set-up fee; n = number of T–shirts ordered. Write a variable expression for the cost of the T–shirts ordered if each T–shirt costs $6 to make.

2. w = wages earned for the year; b = money received as a bonus. Write a variable expression for the total money paid for the year.

3. h = number of hours worked; p = the hourly pay rate. Write a variable expression for how much money you earn.

4. p = hourly earnings; c = money earned for each sale (commission). Write a variable expression for the gross pay an employee would earn who worked 40 hours and made 21 sales.

5. f = number of square feet of lawn. Write a variable expression for the number of bags of No More Bugs Pesticide needed if each bag covers 4,000 square feet.

Cooperative Learning and Algebra 1: Becky Bride
Kagan Publishing • 1 (800) 933-2667 • www.KaganOnline.com

4 EXPRESSIONS IN THE REAL WORLD

2m-4

Structure: Sage-N-Scribe

6. p = the amount of the mortgage payment paid each month. Write a variable expression for the amount of money spent on paying the mortgage for 1 year.

7. b = the balance in your checking account at the beginning of the month; d = the total deposits made during the month; c = the total amount of checks written for the month. Write a variable expression for the balance in your checking account at the end of the month.

8. p = the price of an item. Write a variable expression for the total cost of the item if the sales tax rate is 7%.

9. p = the price of a refrigerator. Write a variable expression for the refrigerator if there is a $200 rebate on the refrigerator and the refrigerator is discounted 15%.

10. u = price of all used music CDs; n = price of all new music CDs. Write a variable expression for the cost of 12 used CDs and 8 new CDs.

Answers on teacher notes (page 97).

LESSON 3
SIMPLIFYING EXPRESSIONS

This lesson begins with an exploratory exercise for students to discover what is meant by "like" and "unlike terms." Reviewing the concept of "area" prior to the investigation is important since algebra tiles are based on the concept of area. The purpose for including this investigation is to reduce the occurrence of adding unlike terms. Included are activities to practice forming expressions using these tiles so the students will be successful with the investigation. Activities processing how to simplify expressions with like terms follow.

Activities requiring students to form more complicated expressions prepare the students for the activity that explores the distributive property. Modeling how to form these expressions is crucial to the success of the investigation. Activities that give students practice forming these expressions are included prior to the investigation so students will be successful with the investigation. Activities using the distributive property are included.

An activity that processes the opposite of an expression is included so students will be able to successfully subtract polynomials later in the school year. It reinforces the concept developed in chapter 1. The activity "Simplify Me—Advanced" puts all the concepts developed in this lesson together for the students to practice.

The final activity is a synthesis activity where each team of students will develop a graphic organizer of this lesson. Not only does it offer review of the concepts, but students can also show how all the pieces they learned fit together.

1 CAN YOU FORM ME WITH TILES?

This activity uses algebra tiles so students see that the representation for 4x and 4 are very different. Students must first understand the concept of area since the physical representation of x, x^2, and constants are based on it. Each unit tile is a square whose side each represents one unit, so its area is 1 square unit. The x tile has a width equal to 1 and a length that is x. So its area is 1x. The x^2 tile is a square whose sides each represent length x. So its area is x^2. In chapter 1, blackline master 1.2.1 contains the tiles for 1, x, and x^2. Blackline master 2.3.1 has tiles for y, y^2, w, xy, and w^2. Modeling with these tiles prior to the exploratory investigation is critical.

▶ **Structure**
 • RallyCoach

▶ **Materials**
 • 1 set of algebra tiles that include 1, x, x^2, y, y^2, w, xy, and w^2 (Blacklines 2.3.1 and 1.2.1)
 • Transparency 2.3.2

Concrete

Setup:
Copy Blackline masters 2.3.1

and 1.2.1 onto red and green cardstock. Cut apart the cards. Red will represent negative quantities and green will represent positive quantities.

1. Teacher poses multiple problems with Transparency 2.3.2.

2. Partner A forms the first expression with algebra tiles.

3. Partner B watches, listens, checks, and coaches.

4. Partner B forms the next expression with algebra tiles.

5. Partner A watches, listens, checks, and coaches.

6. Repeat starting at step 2 for the remaining problems.

ACTIVITY

2 CAN YOU PUT ME TOGETHER?

Modeling expressions containing multiple terms is necessary for this activity, which prepares students for the third activity in this lesson.

Concrete

1. Teacher poses multiple problems with Transparency 2.3.3.

2. Partner A forms the first expression with algebra tiles.

3. Partner B watches, listens, checks, and coaches.

4. Partner B forms the next expression with algebra tiles.

5. Partner A watches, listens, checks, and coaches.

6. Repeat starting at step 2 for the remaining problems.

▶ **Structure**
• RallyCoach

▶ **Materials**
• Algebra tiles
• Transparency 2.3.3

ACTIVITY

3 CAN YOU WRITE MY EXPRESSION?

This activity reverses the process and the students have to write the expression that the algebra tiles model.

Connecting concrete to symbolic

Setup:
In pairs, Student A is the Sage; Student B is the Scribe. Students fold a sheet of paper in half and each writes his/her name on one half.

1. The Sage, using algebra tiles, forms an expression that he/she made up.

2. The Scribe writes symbolically (using variables and numbers) the expression the Sage formed.

3. If the Scribe is correct, the Sage praises the Scribe. Otherwise, the Sage coaches, then praises.

4. Students switch roles for the next problem.

▶ **Structure**
• Sage-N-Scribe

▶ **Materials**
• Algebra tiles
• 1 sheet of paper and pencil per pair

Chapter 2: Expressions
Lesson Three

ACTIVITY
4

EXPLORING LIKE AND UNLIKE TERMS

▶ **Structure**
· Solo-Pair Consensus-Team Consensus

▶ **Materials**
· 1 Blackline 2.3.4 per student
· 1 set of algebra tiles per student
· 1 sheet of graph paper and pen/pencil per student

Exploratory

Solo

1. Have each student complete the investigation individually using the algebra tiles.

Pair Consensus

2. For each problem on the investigation, each student shares with his/her partner, using RallyRobin, his/her response. They discuss the problems that they disagree on, trying to come to consensus on the correct response. For the problems they can't reach consensus on, they mark these so they can focus on them during the team phase. Encourage the students to add to their responses if their partner verbalized an understanding they did not see.

Team Consensus

3. Each pair shares their responses, using RallyRobin, with the other pair in their team, augmenting their responses if necessary. When the teams are through sharing, each student should have a detailed, complete summary.

ACTIVITY
5

FIND MY LIKE TERMS

▶ **Structure**
· RallyCoach

▶ **Materials**
· Transparency 2.3.5

Algebraic

1. Teacher poses many problems using Transparency 2.3.5.

2. In pairs, students take turns stating which terms are like and explaining why they are like terms.

Cooperative Learning and Algebra 1: Becky Bride
Kagan Publishing • 1 (800) 933-2667 • www.KaganOnline.com

6 SIMPLIFY ME WITH TILES

This activity has students simplify expressions with like terms by moving the like terms together, removing any zeros, then stating the resulting equation orally. An example is provided below using the expression: $2x - 3y + 5y - x^2 + 3x$.

Concrete

1. Teacher poses many problems using Transparency 2.3.5.

2. Partner A simplifies the first expression using algebra tiles.

3. Partner B watches, listens, checks, and praises.

4. Partner B simplifies the next problem.

▶ **Structure**
· RallyCoach

▶ **Materials**
· Transparency 2.3.5
· 1 set of algebra tiles per pair

5. Partner A watches and listens, checks, and praises.

6. Repeat for remaining problems starting at step 2.

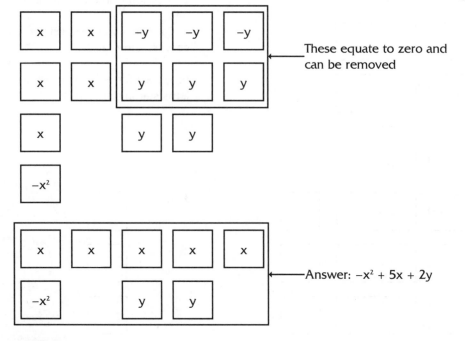

These equate to zero and can be removed

Answer: $-x^2 + 5x + 2y$

ACTIVITY

7 SIMPLIFY ME WITHOUT TILES

Algebraic

Setup:
In pairs, Student A is the Sage; Student B is the Scribe. Students fold a sheet of paper in half and each writes his/her name on one half.

1. The Sage, using algebra tiles, gives the Scribe step-by-

step instructions on how to solve problem 1.

2. The Scribe records the Sage's solution step-by-step in writing.

3. If the Sage is correct, the Scribe praises the Sage. Otherwise, the Scribe coaches, then praises.

▶ **Structure**
· Sage-N-Scribe

▶ **Materials**
· Transparency 2.3.6
· 1 sheet of paper and pencil per pair

4. Students switch roles for the next problem.

Chapter 2: Expressions
Lesson Three

ACTIVITY
8 MAKE ME WITH TILES

▶ **Structure**
· RallyCoach

▶ **Materials**
· Transparency 2.3.7
· 1 set of algebra tiles per pair

This activity and the next prepare the students for the exploratory activity on the distributive property. Modeling with algebra tiles is essential for success on this activity and the next and the exploration following. The students have to be able to show $3 \cdot 2x$ (three groups of $2x$), $3 \cdot 1$ (three groups of 1), and $3(2x + 1)$ (three groups of $2x + 1$) using tiles. Below is an illustration.

Concrete

1. Teacher poses multiple problems with Transparency 2.3.7.

2. Partner A forms the first expression using algebra tiles.

3. Partner B watches, listens, checks, and praises.

4. Partner B forms the next expression using algebra tiles.

5. Partner A watches, listens, checks, and praises.

6. Repeat for remaining problems starting at step 2.

$3 \cdot 2x$

$3(2x + 1)$

$3 \cdot 1$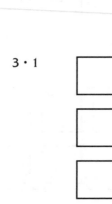

ACTIVITY
9 MAKE ME WITH TILES-ADVANCED

▶ **Structure**
· RallyCoach

▶ **Materials**
· Transparency 2.3.8
· 1 set of algebra tiles per pair

Concrete

1. Teacher poses multiple problems with Transparency 2.3.8.

2. Partner A forms the first expression using algebra tiles.

3. Partner B watches and listens, checks, and praises.

4. Partner B forms the next expression using algebra tiles.

5. Partner A watches and listens, checks, and praises.

6. Repeat for remaining problems starting at step 2.

110

ACTIVITY

10 EXPLORING THE DISTRIBUTIVE PROPERTY

Exploratory

Solo

1. Have each student complete the investigation individually using the algebra tiles.

Pair Consensus

2. For each problem on the investigation, each student shares with his/her partner, using RallyRobin, his/her response. They discuss the problems that they disagree on, trying to come to consensus on the correct response. For the problems they can't

reach consensus on, they mark these so they can focus on them during the team phase. Encourage the students to add to their responses if their partner verbalized an understanding they did not see.

Team Consensus

3. Each pair shares their responses, using RallyRobin, with the other pair in their team, augmenting their responses if necessary. When the teams are through sharing, each student should have a detailed, complete summary.

> ### ▶ Structure
> · Solo-Pair Consensus-Team Consensus
>
> ### ▶ Materials
> · 1 Blackline 2.3.9 per student
> · 1 set of algebra tiles per student
> · 1 sheet of graph paper and pencil per student

ACTIVITY

11 SIMPLIFY ME WITH THE DISTRIBUTIVE PROPERTY

Algebraic

Setup:
In pairs, Student A is the Sage; Student B is the Scribe. Students fold a sheet of paper in half and each writes his/her name on one half.

1. The Sage gives the Scribe step-by-step instructions on how to solve problem 1.

2. The Scribe records the Sage's solution step-by-step in writing.

3. If the Sage is correct, the Scribe praises the Sage. Otherwise, the Scribe coaches, then praises.

4. Students switch roles for the next problem.

> ### ▶ Structure
> · Sage-N-Scribe
>
> ### ▶ Materials
> · Transparency 2.3.10
> · 1 sheet of paper and pencil per pair

Chapter 2: Expressions
Lesson Three

ACTIVITY

12 **FIND MY OPPOSITE**

▶ **Structure**
• Sage-N-Scribe

▶ **Materials**
• Transparency 2.3.11
• 1 sheet of paper and pencil per pair

Algebraic

Setup:
In pairs, Student A is the Sage; Student B is the Scribe. Students fold a sheet of paper in half and each writes his/her name on one half.

1. The Sage gives the Scribe step-by-step instructions on how to solve problem 1.

2. The Scribe records the Sage's instructions step-by-step in writing.

3. If the Sage is correct, the Scribe praises the Sage. Otherwise, the Scribe coaches, then praises.

4. Students switch roles for the next problem.

Note: Modeling with algebra tiles—paying special attention to color—may make this activity go smoother.

ACTIVITY

13 **SIMPLIFY ME-ADVANCED**

▶ **Structure**
• Sage-N-Scribe

▶ **Materials**
• Transparency 2.3.12
• 1 sheet of paper and pencil per pair

Algebraic

Setup:
In pairs, Student A is the Sage; Student B is the Scribe. Students fold a sheet of paper in half and each writes his/her name on one half.

1. The Sage gives the Scribe step-by-step instructions on how to solve problem 1.

2. The Scribe records the Sage's instructions step-by-step in writing.

3. If the Sage is correct, the Scribe praises the Sage. Otherwise, the Scribe coaches, then praises.

4. Students switch roles for the next problem.

ACTIVITY

14 FACTOR ME—DISTRIBUTIVE PROPERTY UNDONE

Algebraic

Setup:
In pairs, Student A is the Sage; Student B is the Scribe. Students fold a sheet of paper in half and each writes his/her name on one half.

1. The Sage gives the Scribe step-by-step instructions on how to solve problem 1.

2. The Scribe records the Sage's solution step-by-step in writing.

3. If the Sage is correct, the Scribe praises the Sage. Otherwise, the Scribe coaches, then praises.

▶ **Structure**
• Sage-N-Scribe

▶ **Materials**
• Transparency 2.3.13
• 1 sheet of paper and pencil per pair

ACTIVITY

15

CREATE AN APPLICATION

This is a fun activity where students will create, as a team effort, applications for expressions. Teammates will define variables and create story problems for their variables through a series of stages. All four teammates begin their own problem. When the process is finished, four applications of expressions have been produced by each team and each teammate has had the opportunity to contribute to each story.

▶ **Structure**
· Simultaneous RoundTable

▶ **Materials**
· 1 sheet of paper and pencil per student

Connect

1. The teacher tells students to select an object to represent a "unit."

2. Each teammate selects something to represent a "unit" for a story problem and writes it at the top of his/her paper. For example, Teammate 1 may write "candy bar" on his/her sheet. Teammate 2 will likely select a different "unit."

3. After the teacher signals time, or students indicate they are all finished by placing their thumbs up, they pass the paper one person clockwise.

4. Each teammate reads the unit his/her teammate wrote and writes a related variable. For the candy bar example, the next student's variable could be "b = the number of candy bars in a bag."

5. When done, students pass their papers one teammate clockwise.

6. Teammates read the unit and variable and add a new related variable. Continuing with our example, the next teammate may write, "c = number of candy bars in a carton."

7. When done, students pass their papers one teammate clockwise.

8. Teammates read the unit and two variables on their new sheet and write a story problem. For example, "Rolando and Gabriel are planning a party. One of the foods they wanted to serve was their favorite candy bars. Rolando bought two bags of candy bars. Not knowing that Rolando had already gone shopping, Gabriel bought five cartons of candy bars. While the boys were not looking, Rolando's little sister took four candy bars. If ten people are invited to the party, how many candy bars does each person receive?"

9. When done, students pass their papers one teammate clockwise.

10. Teammates read the story problem and write an expression to represent the story problem. To continue the candy bar example,

$$\frac{2b + 5c - 4}{12} = \text{number of candy bars per person}$$

11. When done, students pass their papers one teammate clockwise.

12. Students simultaneously check the expression on the paper.

Management Tip: If students use different colored pencils and write their names at the top of each paper, it is easier to hold each individual accountable. As a sponge for this activity, students can write their own story problem.

WHAT DID WE LEARN?

This activity is a summary of the entire expression unit. Encourage the students to reflect back on all they have learned about expressions to make this graphic organizer.

Synthesis via Graphic Organizer

1. Each teammate signs his/her name in the upper-right corner of the team paper with the color pen/pencil he/she is using.

2. One teammate writes "Expressions" in the center of the team paper in a rectangle.

3. Teammate 1 shares with the team one core concept he/she learned in the unit.

4. The student checks for consensus.

5. The teammates show agreement or lack of agreement with thumbs up or down.

6. If there is agreement, the students celebrate and the teammate records the core concept on the graphic organizer, connecting it with a line to the main idea, "Expressions." If not, teammates discuss the response until there is agreement and then they celebrate.

7. Play continues with the next student's core concept until all core concepts are exhausted.

8. Repeat steps 3-7 with teammates, adding details to each core concept and making bridges between related ideas

▶ **Structure**
• RoundTable Consensus

▶ **Materials**
• 1 large sheet of paper per team
• different color pen or pencil for each student on the team

ACTIVITY

1 ALGEBRA TILES

Algebra Tiles: y, y², and w

Cooperative Learning and Algebra 1: Becky Bride
Kagan Publishing • 1 (800) 933-2667 • www.KaganOnline.com

ACTIVITY

1 ALGEBRA TILES

EXPRESSIONS
2m-4

Algebra Tiles: w²

ACTIVITY

1 ALGEBRA TILES

EXPRESSIONS
2m-4

Algebra Tiles: xy

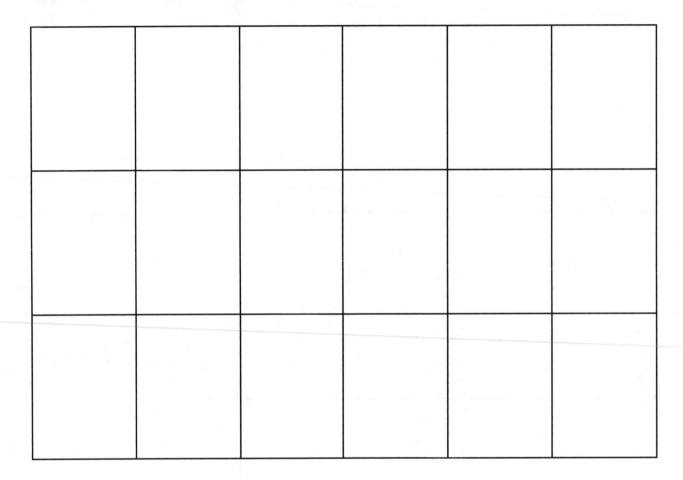

Cooperative Learning and Algebra 1: Becky Bride
Kagan Publishing • 1 (800) 933-2667 • www.KaganOnline.com

1 CAN YOU FORM ME WITH TILES?

Structure: RallyCoach

Form each of the following expressions using algebra tiles.

1. 8

2. –5

3. –3xy

4. 2y

5. $6w^2$

6. $-4y^2$

7. –7

8. 9w

9. $6x^2$

10. –10xy

CAN YOU PUT ME TOGETHER?

Structure: RallyCoach

Form each of the following expressions using algebra tiles.

1. $3xy - 4$

2. $5 + 2y^2$

3. $x^2 - 6w$

4. $5xy + 7y^2$

5. $2 - 6w + xy$

6. $4x + 3y^2 - 1$

7. $7 - w - 8w^2$

8. $2x^2 - 3y + 4w^2$

Cooperative Learning and Algebra 1: Becky Bride
Kagan Publishing • 1 (800) 933-2667 • www.KaganOnline.com

4 EXPLORING LIKE AND UNLIKE TERMS

1. a) Using algebra tiles, build the expression $3x + 4$ and draw a diagram of it on graph paper.

    b) Using algebra tiles, build the expression $7x$ and draw a diagram of it on graph paper.

    c) Compare the tiles for $3x + 4$ with the tiles for $7x$. Is $3x + 4$ the same as $7x$? **EXPLAIN.**

    d) Using algebra tiles, build the expression 7 and draw a diagram of it on graph paper.

    e) Compare the tiles for $3x + 4$ with the tiles for 7. Is $3x + 4$ the same as 7? **EXPLAIN.**

    f) $3x$ and 4 are considered unlike terms. Why are they called "unlike terms"? **EXPLAIN.**

2. a) Using algebra tiles, build the expression $2x^2 + 3x$ and draw a diagram of it on graph paper.

    b) Using algebra tiles, build the expression $5x^2$ and draw a diagram of it on graph paper.

    c) Compare the tiles for $2x^2 + 3x$ with the tiles for $5x^2$. Is $2x^2 + 3x$ the same as $5x^2$? **EXPLAIN.**

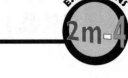

ACTIVITY 4
EXPLORING LIKE AND UNLIKE TERMS

d) Using algebra tiles, build the expression 5x and draw a diagram of it on graph paper.

e) Compare the tiles for $2x^2 + 3x$ with the tiles for 5x. Is $2x^2 + 3x$ the same as 5x? **EXPLAIN.**

f) $2x^2$ and 3x are considered unlike terms. Why are they called "unlike terms"? **EXPLAIN.**

3. a) Using algebra tiles, build the expression 3x + 5xy and draw a diagram of it on graph paper.

    b) Using algebra tiles, build the expression 8xy and draw a diagram of it on graph paper.

    c) Compare the tiles for 3x + 5xy with the tiles for 8xy. Is 3x + 5xy the same as 8xy? **EXPLAIN.**

    d) Using algebra tiles, build the expression 8x and draw a diagram of it on graph paper.

    e) Compare the tiles for 3x + 5xy with the tiles for 8x. Is 3x + 5xy the same as 8x? **EXPLAIN.**

Cooperative Learning and Algebra 1: Becky Bride
Kagan Publishing • 1 (800) 933-2667 • www.KaganOnline.com

EXPLORING LIKE AND UNLIKE TERMS

f) 3x and 5xy are considered unlike terms. Why are they called "unlike terms"? **EXPLAIN.**

4. a) Using algebra tiles, build the expression $x^2 + 3y^2$ and draw a diagram of it on graph paper.

b) Using algebra tiles, build the expression $4x^2$ and draw a diagram of it on graph paper.

c) Compare the tiles for $x^2 + 3y^2$ with the tiles for $4x^2$. Is $x^2 + 3y^2$ the same as $4x^2$? **EXPLAIN.**

d) Using algebra tiles, build the expression $4y^2$ and draw a diagram of it on graph paper.

e) Compare the tiles for $x^2 + 3y^2$ with the tiles for $4y^2$. Is $x^2 + 3y^2$ the same as $4y^2$? **EXPLAIN.**

f) x^2 and $3y^2$ are considered unlike terms. Why are they called "unlike terms"? **EXPLAIN.**

ACTIVITY
4

EXPLORING LIKE AND UNLIKE TERMS

5. a) Using algebra tiles, build the expression $4x^2 + 3x^2$ and draw a diagram of it on graph paper.

   b) Using algebra tiles, build the expression $7x^2$ and draw a diagram of it on graph paper.

   c) Compare the tiles for $4x^2 + 3x^2$ with the tiles for $7x^2$. Is $4x^2 + 3x^2$ the same as $7x^2$? **EXPLAIN.**

   d) $4x^2$ and $3x^2$ are considered like terms. Why are they called "like terms"? **EXPLAIN.**

6. a) Using algebra tiles, build the expression $2xy + 3xy$ and draw a diagram of it on graph paper.

   b) Using algebra tiles, build the expression $5xy$ and draw a diagram of it on graph paper.

   c) Compare the tiles for $2xy + 3xy$ with the tiles for $5xy$. Is $2xy + 3xy$ the same as $5xy$? **EXPLAIN.**

   d) $2xy$ and $3xy$ are considered like terms. Why are they called "like terms"? **EXPLAIN.**

Cooperative Learning and Algebra 1: Becky Bride
Kagan Publishing • 1 (800) 933-2667 • www.KaganOnline.com

EXPLORING LIKE AND UNLIKE TERMS

EXPRESSIONS
2m-4

7. a) Using algebra tiles, build the expression 6x + (–2x) and draw a diagram of it on graph paper.

    b) Using algebra tiles, build the expression 4x and draw a diagram of it on graph paper.

    c) Compare the tiles for 6x + (–2x) (after the zeros are removed) with the tiles for 4x. Is 6x + (–2x) the same as 4x? **EXPLAIN.**

    d) 6x and (–2x) are considered like terms. Why are they called "like terms"? **EXPLAIN.**

8. Summarize the investigation. Include a description of like terms and unlike terms and the type of terms that can be simplified.

9. Rewrite $2x + 3y - 5x^2 + 2y - x$, simplifying like terms.

FIND MY LIKE TERMS
SIMPLIFY ME WITH TILES

Structure: RallyCoach

Activity 5: Identify the like terms in each problem; then explain why they are like terms.
Activity 6: Simplify by using algebra tiles.

1. $4x + 2y - 3x + 4 - y$

2. $5x^2 - 2x + 8 - 8x + 1$

3. $7x + 2xy - 2x^2 + 5xy - 3$

4. $-3x^2 + 3y^2 - 5x + 4y^2$

5. $9w^2 - 6y^2 + 2xy - 7w^2 - 6xy$

6. $12 - 4x + 4x^2 - 7 - 8x^2$

Activity 6 Answers:
1. $x + y + 4$
2. $5x^2 - 10x + 9$
3. $7x + 7xy - 2x^2 - 3$
4. $-3x^2 + 7y^2 - 5x$
5. $2w^2 - 6y^2 - 4xy$
6. $5 - 4x - 4x^2$

Cooperative Learning and Algebra 1: Becky Bride
Kagan Publishing • 1 (800) 933-2667 • www.KaganOnline.com

SIMPLIFY ME WITHOUT TILES

Structure: Sage-N-Scribe

Simplify each expression.

1. $4w + 6k - 3w + 2 - k$

2. $5x^4 - 2x + 8 - 9x^4 + 1x$

3. $7c + 2cf - 2f^2 + 3cf - 3c$

4. $-3w^2 + x^2 - 5w + 4x^2$

5. $9d^2 - y^2 + 2xy - d^2 - 6xy$

6. $12 - x + 4x^2 - 7x - 5x^2$

7. $5m^3 - 4mp^2 + m^2p - 2m^3 + 6mp^2$

8. $2g - 5gh + 1 + 3hg - g - h$

Answers:
1. $5k + w + 2$
2. $-4x^4 - x + 8$
3. $4c + 5cf - 2f^2$
4. $-3w^2 + 5x^2 - 5w$
5. $8d^2 - y^2 - 4xy$
6. $12 - 8x - x^2$
7. $3m^3 + 2mp^2 + m^2p$
8. $-2gh + g - h + 1$

EXPRESSIONS
2m-4

ACTIVITY 8

MAKE ME WITH TILES

Structure: RallyCoach

Form each expression using algebra tiles.

1. 5 · 2

2. 3 · 4x

3. 2 · w

4. 3 · (−2)

5. 2 · (−3y)

6. 3 · (x²)

7. 4 · (−2xy)

8. 2 · 5w

Cooperative Learning and Algebra 1: Becky Bride
Kagan Publishing • 1 (800) 933-2667 • www.KaganOnline.com

MAKE ME WITH TILES—ADVANCED

Structure: RallyCoach

Form each expression using algebra tiles.

1. $2(3x - 1)$

2. $3(x + 2)$

3. $3(x^2 + 4)$

4. $2(2y^2 - w)$

5. $4(w - 2y - 3)$

6. $3(x + y + w^2)$

7. $2(3y - y^2 + 2x)$

8. $5(x + 2xy - 2y)$

ACTIVITY
10

EXPLORING THE DISTRIBUTIVE PROPERTY

EXPRESSIONS
2m-4

1. a) Using algebra tiles, build the expression $2(3x + 4)$ and draw a diagram of the tiles on graph paper.

    b) Using algebra tiles, build the expression $2 \cdot 3x + 4$ and draw a diagram of the tiles on graph paper.

    c) Compare the tiles for $2(3x + 4)$ with the tiles for $2 \cdot 3x + 4$. Is $2(3x + 4)$ the same as $2 \cdot 3x + 4$? **EXPLAIN.**

    d) Using algebra tiles, build the expression $3x + 2 \cdot 4$ and draw a diagram of the tiles on graph paper.

    e) Compare the tiles for $2(3x + 4)$ with the tiles for $3x + 2 \cdot 4$. Is $2(3x + 4)$ the same as $3x + 2 \cdot 4$? **EXPLAIN.**

    f) Using algebra tiles, build the expression $2 \cdot 3x + 2 \cdot 4$ and draw a diagram of the tiles on graph paper.

    g) Compare the tiles for $2(3x + 4)$ with the tiles for $2 \cdot 3x + 2 \cdot 4$. Is $2(3x + 4)$ the same as $2 \cdot 3x + 2 \cdot 4$? **EXPLAIN.**

2. a) Using algebra tiles, build the expression $3(x^2 + 4y)$ and draw a diagram of the tiles on graph paper.

Cooperative Learning and Algebra 1: Becky Bride
Kagan Publishing • 1 (800) 933-2667 • www.KaganOnline.com

EXPLORING THE
DISTRIBUTIVE PROPERTY

b) Using algebra tiles, build the expression $3 \cdot x^2 + 4y$ and draw a diagram of the tiles on graph paper.

c) Compare the tiles for $3(x^2 + 4y)$ with the tiles for $3 \cdot x^2 + 4y$. Is $3(x^2 + 4y)$ the same as $3 \cdot x^2 + 4y$? **EXPLAIN.**

d) Using algebra tiles, build the expression $x^2 + 3 \cdot 4y$ and draw a diagram of the tiles on graph paper.

e) Compare the tiles for $3(x^2 + 4y)$ with the tiles for $x^2 + 3 \cdot 4y$. Is $3(x^2 + 4y)$ the same as $x^2 + 3 \cdot 4y$? **EXPLAIN.**

f) Using algebra tiles, build the expression $3 \cdot x^2 + 3 \cdot 4y$ and draw a diagram of the tiles on graph paper.

g) Compare the tiles for $3(x^2 + 4y)$ with the tiles for $3 \cdot x^2 + 3 \cdot 4y$. Is $3(x^2 + 4y)$ the same as $3 \cdot x^2 + 3 \cdot 4y$? **EXPLAIN.**

3. a) Using algebra tiles, build the expression $3(2x^2 + 3xy - 1)$ and draw a diagram of the tiles on graph paper.

    b) Using algebra tiles, build the expression $3 \cdot 2x^2 + 3xy - 1$ and draw a diagram of the tiles on graph paper.

EXPLORING THE DISTRIBUTIVE PROPERTY

EXPRESSIONS
2m-4

c) Compare the tiles for $3(2x^2 + 3xy - 1)$ with the tiles for $3 \cdot 2x^2 + 3xy - 1$. Is $3(2x^2 + 3xy - 1)$ the same as $3 \cdot 2x^2 + 3xy - 1$? **EXPLAIN.**

d) Using algebra tiles, build the expression $2x^2 + 3 \cdot 3xy - 1$ and draw a diagram of the tiles on graph paper.

e) Compare the tiles for $3(2x^2 + 3xy - 1)$ with the tiles for $2x^2 + 3 \cdot 3xy - 1$. Is $3(2x^2 + 3xy - 1)$ the same as $2x^2 + 3 \cdot 3xy - 1$? **EXPLAIN.**

f) Using algebra tiles, build the expression $2x^2 + 3xy - 3 \cdot 1$ and draw a diagram of the tiles on graph paper.

g) Compare the tiles for $3(2x^2 + 3xy - 1)$ with the tiles for $2x^2 + 3xy - 3 \cdot 1$. Is $3(2x^2 + 3xy - 1)$ the same as $2x^2 + 3xy - 3 \cdot 1$? **EXPLAIN.**

h) Using algebra tiles, build the expression $3 \cdot 2x^2 + 3 \cdot 3xy - 3 \cdot 1$ and draw a diagram of the tiles on graph paper.

i) Compare the tiles for $3(2x^2 + 3xy - 1)$ with the tiles for $3 \cdot 2x^2 + 3 \cdot 3xy - 3 \cdot 1$. Is $3(2x^2 + 3xy - 1)$ the same as $3 \cdot 2x^2 + 3 \cdot 3xy - 3 \cdot 1$? **EXPLAIN.**

Cooperative Learning and Algebra 1: Becky Bride
Kagan Publishing • 1 (800) 933-2667 • www.KaganOnline.com

ACTIVITY 10 EXPLORING THE DISTRIBUTIVE PROPERTY

4. What does the word "distribute" mean?

5. a) Refer back to problem 1. Which expression was equivalent to $2(3x + 4)$? Write it below.

   b) Refer back to problem 2. Which expression was equivalent to $3(x^2 + 4y)$? Write it below.

   c) Refer back to problem 3. Which expression was equivalent to $3(2x^2 + 3xy - 1)$? Write it below.

6. Compare the equivalent expressions in problem 5. All of these are examples of the distributive property. Based on your observations in this investigation, rewrite each of the following using the distributive property.

   a) $-5(4x + 7)$

   b) $3(2x^3 - 4x^2 + 5)$

   c) $8(3x^2 - xy - 6)$

SIMPLIFY ME WITH THE DISTRIBUTIVE PROPERTY

EXPRESSIONS

2m-4

Structure: Sage-N-Scribe

Simplify each expression.

1. $3(4x - 5y)$

2. $2(3x^2 + 2x - 7)$

3. $-7(3y^2 + 4y - 2)$

4. $4(9x + 3xy + 6y)$

5. $-3(4x^2 - 3x - 5)$

6. $-6(3y^2 + 2x - w)$

Answers:
1. $12x - 15y$
2. $6x^2 + 4x - 14$
3. $-21y^2 - 28y + 14$
4. $36x + 12xy + 24y$
5. $-12x^2 + 9x + 15$
6. $-18y^2 - 12x + 6w$

Cooperative Learning and Algebra 1: Becky Bride
Kagan Publishing • 1 (800) 933-2667 • www.KaganOnline.com

12 FIND MY OPPOSITE

Structure: Sage-N-Scribe

Simplify each expression.

1. $-(3x - 4p + 7)$

2. $-(c^4 - 3h^3 - 5)$

3. $-(5f^2 + 8g - 9dp)$

4. $-(2rt - y + 6)$

5. $-(12v^2 + 6w - 7d^2)$

6. $-(5x^2y - 11 - 3xy^2)$

7. $-(-f^2 - 3ab^4 + 10a^2b^7)$

8. $-(-e^3 + 14g - 7eg^4)$

Answers:
1. $-3x + 4p - 7$
2. $-c^4 + 3h^3 + 5$
3. $-5f^2 - 8g + 9dp$
4. $-2rt + y - 6$
5. $-12v^2 - 6w + 7d^2$
6. $-5x^2y + 11 + 3xy^2$
7. $f^2 + 3ab^4 - 10a^2b^7$
8. $e^3 - 14g + 7eg^4$

13 SIMPLIFY ME—ADVANCED

Structure: Sage-N-Scribe

Simplify each expression.

1. $4x + 3(2x - 6y) - y$

2. $5x^2 - 2(3x^2 - xy) + 4xy - 8$

3. $-2(8m + 5p) - (6p - 3m)$

4. $-(4x^2 - 3x + 1) + 3(4 - 5x)$

5. $7x^3 - 8x^2 + 3(5x^3 - x^2) + 2$

6. $5m - (4m^2 + 6m - 2) + 3(5m + 3)$

Answers:
1. $10x - 19y$
2. $-x^2 + 6xy - 8$
3. $-13m - 16p$
4. $-4x^2 - 12x + 11$
5. $22x^3 - 11x^2 + 2$
6. $-4m^2 + 14m + 11$

Cooperative Learning and Algebra 1: Becky Bride
Kagan Publishing • 1 (800) 933-2667 • www.KaganOnline.com

FACTOR ME—DISTRIBUTIVE PROPERTY UNDONE

EXPRESSIONS 2m-4

Structure: Sage-N-Scribe

Factor each expression.

1. $12x - 4xy + 8$

2. $-9y + 12xy - 6x$

3. $24x^3 - 3x^2 + 9$

4. $50h^3 - 40h^2x^2 - 30x^2$

5. $-14w^4 + 21x^5$

6. $32d^3 - 24p^4$

Answers:
1. $4(3x - xy + 2)$
2. $3(-3y + 4xy - 2x)$
3. $3(8x^3 - x^2 + 3)$
4. $10(5h^3 - 4h^2x^2 - 3x^2)$
5. $7(-2w^4 + 3x^5)$
6. $8(4d^3 - 3p^4)$

LESSON 4
EXPONENTS

This lesson focuses on exponents and the rules of exponents. Activities 1 and 2 have student work with the definition of an "exponent" that will be important in the exploration activities. After each exploration, there is an activity to process that rule. Negative and zero exponents are included. The final activity has the students synthesize what they have learned.

ACTIVITY

1 EXPAND ME

▶ **Structure**
· RallyCoach

▶ **Materials**
· Transparency 2.4.1
· 1 sheet of paper and pencil per pair

Algebraic

1. Teacher poses multiple problems with Transparency 2.4.1.

2. Partner A expands the expression in problem 1.

3. Partner B watches and listens, checks, and praises.

4. Partner B expands the next problem.

5. Partner A watches and listens, checks, and praises.

6. Repeat with remaining problems starting at step 2.

ACTIVITY

2 REWRITE ME

▶ **Structure**
· RallyCoach

▶ **Materials**
· Transparency 2.4.2
· 1 sheet of paper and pencil per pair

Algebraic

1. Teacher poses multiple problems with Transparency 2.4.2.

2. Partner A rewrites problem 1 using exponents.

3. Partner B watches and listens, checks, and praises.

4. Partner B rewrites the next problem.

5. Partner A watches and listens, checks, and praises.

6. Repeat with remaining problems, starting at step 2.

Chapter 2: Expressions
Lesson Four

138

EXPLORING THE PRODUCT RULE OF EXPONENTS

Exploratory

Solo

1. Have each student complete the investigation individually.

Pair Consensus

2. For each problem on the investigation, each student shares with his/her partner, using RallyRobin, his/her response. They discuss the problems that they disagree on trying to come to consensus on the correct response. For the problems they can't

reach consensus on, they mark these so they can focus on them during the team phase. Encourage the students to add to their responses if their partner verbalized an understanding they did not see.

Team Consensus

3. Each pair shares their responses, using RallyRobin, with the other pair in their team, augmenting their responses if necessary. When the teams are through sharing, each student should have a detailed, complete summary.

▶ **Structure**
· Solo-Pair Consensus-Team Consensus

▶ **Materials**
· 1 Blackline 2.4.3 per student
· 1 sheet of paper and pen/ pencil per student

USE THE PRODUCT RULE

Algebraic

Setup:
In pairs, Student A is the Sage; Student B is the Scribe. Students fold a sheet of paper in half and each writes his/her name on one half.

1. The Sage tells the Scribe how to simplify problem 1.

2. The Scribe records the Sage's instructions step-by-step in writing.

3. If the Sage is correct, the Scribe praises the Sage. Otherwise, the Scribe coaches, then praises.

4. Students switch roles for the next problem.

▶ **Structure**
· Sage-N-Scribe

▶ **Materials**
· Transparency 2.4.4
· 1 sheet of paper and pencil per pair

ACTIVITY 5
EXPLORING THE QUOTIENT RULE OF EXPONENTS

A discussion on factors of 1 (e.g. $\frac{2}{2}$, $\frac{w}{w}$, $\frac{6xy}{6xy}$) prior to this activity will review the prerequisite concept necessary for successful implementation of this activity.

▶ **Structure**
· Solo-Pair Consensus-Team Consensus

▶ **Materials**
· Blackline 2.4.5
· 1 sheet of paper and pen/pencil per student

Exploratory

Solo

1. Have each student complete the investigation individually.

Pair Consensus

2. For each problem on the investigation, each student shares with his/her partner, using RallyRobin, his/her response. They discuss the problems that they disagree on, trying to come to consensus on the correct response. For the problems they can't reach consensus on, they mark these so they can focus on them during the team phase. Encourage the students to add to their responses if their partner verbalized an understanding they did not see.

Team Consensus

3. Each pair shares their responses, using RallyRobin, with the other pair in their team, augmenting their responses if necessary. When the teams are through sharing, each student should have a detailed, complete summary.

ACTIVITY 6
USE THE QUOTIENT RULE

▶ **Structure**
· Sage-N-Scribe

▶ **Materials**
· Transparency 2.4.6
· 1 sheet of paper and pen/pencil per student

Algebraic

Setup:
In pairs, Student A is the Sage; Student B is the Scribe. Students fold a sheet of paper in half and each writes his/her name on one half.

1. The Sage tells the Scribe how to simplify problem 1.

2. The Scribe records the Sage's instructions step-by-step in writing.

3. If the Sage is correct, the Scribe praises the Sage. Otherwise, the Scribe coaches, then praises.

4. Students switch roles for the next problem.

ACTIVITY 7
EXPLORING THE POWER OF A PRODUCT/QUOTIENT RULE

Exploratory

Solo

1. Have each student complete the investigation individually.

Pair Consensus

2. For each problem on the investigation, each student shares with his/her partner, using RallyRobin, his/her response. They discuss the problems that they disagree on, trying to come to consensus on the correct response. For the problems they can't reach consensus on, they mark these so they can focus on them during the team phase. Encourage the students to add to their responses if their partner verbalized an understanding they did not see.

Team Consensus

3. Each pair shares their responses, using RallyRobin, with the other pair in their team, augmenting their responses if necessary. When the teams are through sharing, each student should have a detailed, complete summary.

▶ **Structure**
· Solo-Pair Consensus-Team Consensus

▶ **Materials**
· 1 Blackline 2.4.7 per student
· 1 sheet of paper and pen/pencil per student

ACTIVITY 8
USE THE POWER OF A PRODUCT/QUOTIENT RULE

Algebraic

Setup:
In pairs, Student A is the Sage; Student B is the Scribe. Students fold a sheet of paper in half and each writes his/her name on one half.

1. The Sage tells the Scribe how to simplify problem 1.

2. The Scribe records the Sage's instructions step-by-step in writing.

3. If the Sage is correct, the Scribe praises the Sage. Otherwise, the Scribe coaches, then praises.

4. Students switch roles for the next problem.

▶ **Structure**
· Sage-N-Scribe

▶ **Materials**
· Transparency 2.4.8
· 1 sheet of paper and pencil per pair

ACTIVITY
9 EXPLORING NEGATIVE AND ZERO EXPONENTS

▶ **Structure**
· Solo-Pair Consensus-Team Consensus

▶ **Materials**
· 1 Blackline 2.4.9 per student
· 1 sheet of paper and pen/pencil per student

Exploratory

Solo

1. Have each student complete the investigation individually.

Pair Consensus

2. For each problem on the investigation, each student shares with his/her partner, using RallyRobin, his/her response. They discuss the problems that they disagree on, trying to come to consensus on the correct response. For the problems they can't

reach consensus on, they mark these so they can focus on them during the team phase. Encourage the students to add to their responses if their partner verbalized an understanding they did not see.

Team Consensus

3. Each pair shares their responses, using RallyRobin, with the other pair in their team, augmenting their responses if necessary. When the teams are through sharing, each student should have a detailed, complete summary.

ACTIVITY
10 REWRITE ME WITH POSITIVE EXPONENTS

▶ **Structure**
· Sage-N-Scribe

▶ **Materials**
· Transparency 2.4.10
· 1 sheet of paper and pencil per pair

Algebraic

Setup:
In pairs, Student A is the Sage; Student B is the Scribe. Students fold a sheet of paper in half and each writes his/her name on one half.

1. The Sage tells the Scribe how to rewrite problem 1, using positive exponents.

2. The Scribe records the Sage's instructions step-by-step in writing.

3. If the Sage is correct, the Scribe praises the Sage. Otherwise, the Scribe coaches, then praises.

4. Students switch roles for the next problem.

ACTIVITY

11

DISTRIBUTE PROPERTY REVISITED

Algebraic

Setup:
In pairs, Student A is the Sage; Student B is the Scribe. Students fold a sheet of paper in half and each writes his/her name on one half.

1. The Sage tells the Scribe how to simplify problem 1.

2. The Scribe records the Sage's instructions step-by-step in writing.

3. If the Sage is correct, the Scribe praises the Sage. Otherwise, the Scribe coaches, then praises.

4. Students switch roles for the next problem.

▶ **Structure**
• Sage-N-Scribe

▶ **Materials**
• Transparency 2.4.11
• 1 sheet of paper and pencil per pair

ACTIVITY
12

FACTOR ME—REVISITED

Algebraic

Setup:
In pairs, Student A is the Sage; Student B is the Scribe. Students fold a sheet of paper in half and each writes his/her name on one half.

1. The Sage tells the Scribe how to factor problem 1.

2. The Scribe records the Sage's instructions step-by-step in writing.

3. If the Sage is correct, the Scribe praises the Sage. Otherwise, the Scribe coaches, then praises.

4. Students switch roles for the next problem.

▶ **Structure**
• Sage-N-Scribe

▶ **Materials**
• Transparency 2.4.12
• 1 sheet of paper and pencil per pair

ACTIVITY

13

WHAT DID WE LEARN?

▶ **Structure**
• RoundTable Consensus

▶ **Materials**
• 1 large sheet of paper per team
• different color pen or pencil for each student on the team

Synthesis via Graphic Organizer

1. Each teammate signs his/her name in the upper-right corner of the team paper with the color pen/pencil he/she is using.

2. One teammate writes "Exponents" in the center of the team paper in a rectangle.

3. Teammate 1 shares with the team one core concept he/she learned in the unit.

4. The student checks for consensus.

5. The teammates show agreement or lack of agreement with thumbs up or down.

6. If there is agreement, the students celebrate and the teammate records the core concept on the graphic organizer, connecting it with a line to the main idea, "Exponents." If not, teammates discuss the response until there is agreement and then they celebrate.

7. Play continues with the next student's core concept until all core concepts are exhausted.

8. Repeat steps 3-7, with teammates adding details to each core concept and making bridges between related ideas.

Cooperative Learning and Algebra 1: Becky Bride
Kagan Publishing • 1 (800) 933-2667 • www.KaganOnline.com

ACTIVITY 1 EXPAND ME

Structure: RallyCoach

Using the definition of an "exponent," expand each of the following.

1. 3^8

2. x^5y^2

3. $(4x)^3(2y)^6$

4. $(x + 3)^4$

5. $\dfrac{3^2}{5^3}$

6. $(5xy)^6$

7. $(8x - 1)^3$

8. $\left(\dfrac{2}{3}\right)^5$

Answers:
1. $3 \cdot 3 \cdot 3 \cdot 3 \cdot 3 \cdot 3 \cdot 3 \cdot 3$
2. $x \cdot x \cdot x \cdot x \cdot x \cdot y \cdot y$
3. $(4x \cdot 4x \cdot 4x)(2y \cdot 2y \cdot 2y \cdot 2y \cdot 2y \cdot 2y)$
4. $(x + 3)(x + 3)(x + 3)(x + 3)$
5. $\dfrac{3 \cdot 3}{5 \cdot 5 \cdot 5}$
6. $(5xy)(5xy)(5xy)(5xy)(5xy)(5xy)$
7. $(8x - 1)(8x - 1)(8x - 1)$
8. $\dfrac{2}{3} \cdot \dfrac{2}{3} \cdot \dfrac{2}{3} \cdot \dfrac{2}{3} \cdot \dfrac{2}{3}$

ACTIVITY 2 · REWRITE ME

Structure: RallyCoach

EXPRESSIONS
2m-4

Rewrite each expression using exponents.

1. $4 \cdot 4 \cdot 4 \cdot 4 \cdot 4 \cdot 4 \cdot 4$

2. $(-7x)(-7x)(-7x)$

3. $(x + 1)(x + 1)(x + 1)(x + 1)(x + 1)(x + 1)$

4. $\left(\dfrac{3}{7}\right)\left(\dfrac{3}{7}\right)\left(\dfrac{3}{7}\right)\left(\dfrac{3}{7}\right)$

5. $w \cdot w \cdot w \cdot p \cdot p \cdot p \cdot p$

6. $3 \cdot 3 \cdot 3 \cdot 3 \cdot 3 \cdot w \cdot w \cdot w \cdot x \cdot x \cdot y$

Answers:
1. 4^7
2. $(-7x)^3$
3. $(x + 1)^6$
4. $\left(\dfrac{3}{7}\right)^4$
5. w^3p^4
6. $3^5w^3x^2y$

Cooperative Learning and Algebra 1: Becky Bride
Kagan Publishing • 1 (800) 933-2667 • www.KaganOnline.com

EXPLORING THE PRODUCT RULE OF EXPONENTS

1. $(x^3)(x^2)$

   a) Expand the expression using the definition of an "exponent."

   b) Rewrite the expanded form using exponents.

   c) Compare the exponent in part b to the exponents in the original problem. What operation—add, subtract, multiply, or divide—can be used on the exponents in the original problem to get the exponent in part b?

2. $(w^3)(w^4)$

   a) Expand the expression using the definition of an "exponent."

   b) Rewrite the expanded form using exponents.

   c) Compare the exponent in part b to the exponents in the original problem. What operation—add, subtract, multiply, or divide—can be used on the exponents in the original problem to get the exponent in part b?

3. $(h^2)(h^5)$

   a) Expand the expression using the definition of an "exponent."

   b) Rewrite the expanded form using exponents.

EXPLORING THE PRODUCT RULE OF EXPONENTS

c) Compare the exponent in part b to the exponents in the original problem. What operation—add, subtract, multiply, or divide—can be used on the exponents in the original problem to get the exponent in part b?

4. $(p^4)(p)$

a) Expand the expression using the definition of an "exponent."

b) Rewrite the expanded form using exponents.

c) Compare the exponent in part b to the exponents in the original problem. What operation—add, subtract, multiply, or divide—can be used on the exponents in the original problem to get the exponent in part b?

5. $(n^3)(n^5)$

a) Expand the expression using the definition of an "exponent."

b) Rewrite the expanded form using exponents.

c) Compare the exponent in part b to the exponents in the original problem. What operation—add, subtract, multiply, or divide—can be used on the exponents in the original problem to get the exponent in part b?

Cooperative Learning and Algebra 1: Becky Bride
Kagan Publishing • 1 (800) 933-2667 • www.KaganOnline.com

EXPLORING THE PRODUCT RULE OF EXPONENTS

EXPRESSIONS
2m-4

6. $(3^2x^4y^2)(3xy^5)$

a) Expand the expression using the definition of an "exponent."

b) Rewrite the expanded form using exponents.

c) Compare the exponent in part b to the exponents in the original problem. What operation—add, subtract, multiply, or divide—can be used on the exponents in the original problem to get the exponent in part b?

7. You have just explored the Product Rule for Exponents. Write in your own words what that rule says.

For problems 8-10, use the rule you wrote in problem 7 to simplify the expressions below.

8. $(d^{21})(d^{45})$

9. $(k^{13}g^9)(k^{33}g^{42})$

10. $(2x^2w^5)(23x^{15}w^{31})$

4 USE THE PRODUCT RULE

Structure: Sage-N-Scribe

Simplify each expression.

1. $(w^4)(w^{14})$

2. $(p^8)(p^5)$

3. $(3x^4y^7w^9)(4x^5yw^{10})$

4. $(-5h^6pk^{11})(8h^3p^6k^{21})$

5. $(j^{2x+3})(j^{3x-4})$

6. $(g^{-5w+1})(g^{3w+5})$

Answers:
1. w^{18}
2. p^{13}
3. $12x^9y^8w^{19}$
4. $-40h^9p^7k^{32}$
5. j^{5x-1}
6. g^{-2w+6}

Cooperative Learning and Algebra 1: Becky Bride
Kagan Publishing • 1 (800) 933-2667 • www.KaganOnline.com

ACTIVITY 5

EXPLORING THE QUOTIENT RULE OF EXPONENTS

1. $\dfrac{w^3}{w^2}$

    a) Expand the expression above using the definition of an "exponent."

    b) Remove factors of 1 and rewrite the simplified expression using exponents.

    c) Compare the exponent in part b to the exponents in the original problem. What operation—add, subtract, multiply, or divide—can be used on the exponents in the original problem to get the exponent in part b?

2. $\dfrac{h^6}{h^8}$

    a) Expand the expression above using the definition of an "exponent."

    b) Remove factors of 1 and rewrite the simplified expression using exponents.

    c) Compare the exponent in part b to the exponents in the original problem. What operation—add, subtract, multiply, or divide—can be used on the exponents in the original problem to get the exponent in part b?

EXPLORING THE QUOTIENT RULE OF EXPONENTS

EXPRESSIONS
2m–4

3. $\dfrac{d^5}{d^8}$

    a) Expand the expression above using the definition of an "exponent."

    b) Remove factors of 1 and rewrite the simplified expression using exponents.

    c) Compare the exponent in part b to the exponents in the original problem. What operation—add, subtract, multiply, or divide—can be used on the exponents in the original problem to get the exponent in part b?

4. $\dfrac{k^8}{k^3}$

    a) Expand the expression above using the definition of an "exponent."

    b) Remove factors of 1 and rewrite the simplified expression using exponents.

    c) Compare the exponent in part b to the exponents in the original problem. What operation—add, subtract, multiply, or divide—can be used on the exponents in the original problem to get the exponent in part b?

Cooperative Learning and Algebra 1: Becky Bride
Kagan Publishing • 1 (800) 933-2667 • www.KaganOnline.com

EXPLORING THE QUOTIENT RULE OF EXPONENTS

EXPRESSIONS
$2m-4$

5. $\dfrac{m^6}{m^3}$

a) Expand the expression above using the definition of an "exponent."

b) Remove factors of 1 and rewrite the simplified expression using exponents.

c) Compare the exponent in part b to the exponents in the original problem. What operation—add, subtract, multiply, or divide—can be used on the exponents in the original problem to get the exponent in part b?

6. $\dfrac{g^9 h^5}{g^4 h^3}$

a) Expand the expression above using the definition of an "exponent."

b) Remove factors of 1 and rewrite the simplified expression using exponents.

c) Compare the exponents in part b to the exponents in the original problem. What operation—add, subtract, multiply, or divide—can be used on the exponents in the original problem to get the exponents in part b?

EXPLORING THE QUOTIENT RULE OF EXPONENTS

7. You have just explored the Quotient Rule for Exponents. Write in your own words what that rule says.

For problems 8-10, use your rule from problem 7 to simplify the expressions below.

8. $\dfrac{g^{45}}{g^{27}}$

9. $\dfrac{k^{18}w^{27}}{k^{23}w^{15}}$

10. $\dfrac{m^{24}v^{12}}{m^{13}v}$

Cooperative Learning and Algebra 1: Becky Bride
Kagan Publishing • 1 (800) 933-2667 • www.KaganOnline.com

ACTIVITY

6 USE THE QUOTIENT RULE

Structure: Sage-N-Scribe

Simplify each expression showing required work. All answers must have positive exponents.

1. $\dfrac{g^{23}}{g^{12}}$

2. $\dfrac{h^{12}}{h^{18}}$

3. $\dfrac{12w^9h^4}{10h^{11}w^3}$

4. $\dfrac{32d^{13}p^{14}}{18d^8p^6}$

Answers:

1. g^{11}

2. $\dfrac{1}{h^6}$

3. $\dfrac{6w^6}{5h^7}$

4. $\dfrac{16d^5p^8}{9}$

ACTIVITY 7
EXPLORING THE POWER OF A PRODUCT/QUOTIENT RULE

1. $(w^3)^4$

   a) Expand the expression above using the definition of an "exponent."

   b) Rewrite the expression using exponents.

   c) Compare the exponent in part b to the exponents in the original problem. What operation—add, subtract, multiply, or divide—can be used on the exponents in the original problem to get the exponent in part b?

2. $(w^2)^3$

   a) Expand the expression above using the definition of an "exponent."

   b) Rewrite the expression using exponents.

   c) Compare the exponent in part b to the exponents in the original problem. What operation—add, subtract, multiply, or divide—can be used on the exponents in the original problem to get the exponent in part b?

Cooperative Learning and Algebra 1: Becky Bride
Kagan Publishing • 1 (800) 933-2667 • www.KaganOnline.com

EXPLORING THE POWER OF A PRODUCT/QUOTIENT RULE

EXPRESSIONS

2m-4

3. $(h^2t)^5$

a) Expand the expression above using the definition of an "exponent."

b) Rewrite the expression using exponents.

c) Compare the exponents in part b to the exponents in the original problem. What operation—add, subtract, multiply, or divide—can be used on the exponents in the original problem to get the exponents in part b?

4. $\left(\dfrac{h^2}{p^3}\right)^2$

a) Expand the expression above using the definition of an "exponent."

b) Rewrite the expression using exponents.

c) Compare the exponents in part b to the exponents in the original problem. What operation—add, subtract, multiply, or divide—can be used on the exponents in the original problem to get the exponents in part b?

ACTIVITY 7

EXPLORING THE POWER OF A PRODUCT/QUOTIENT RULE

5. $\left(\dfrac{d^2}{m}\right)^3$

   a) Expand the expression above using the definition of an "exponent."

   b) Rewrite the expression using exponents.

   c) Compare the exponents in part b to the exponents in the original problem. What operation—add, subtract, multiply, or divide—can be used on the exponents in the original problem to get the exponents in part b?

6. $\left(\dfrac{b^3}{w^4}\right)^3$

   a) Expand the expression above using the definition of an "exponent."

   b) Rewrite the expression using exponents.

   c) Compare the exponents in part b to the exponents in the original problem. What operation—add, subtract, multiply, or divide—can be used on the exponents in the original problem to get the exponents in part b?

Cooperative Learning and Algebra 1: Becky Bride
Kagan Publishing • 1 (800) 933-2667 • www.KaganOnline.com

EXPLORING THE POWER OF A PRODUCT/QUOTIENT RULE

EXPRESSIONS
2m-4

7. You have just explored the Power of a Product/Quotient Rule for Exponents. Write in your own words what that rule says.

For problems 8-10, use your rule from problem 7 to simplify the expressions below without expanding them.

8. $(n^{21})^5$

9. $(c^{13}u^{15})^4$

10. $\left(\dfrac{f^{42}}{y^{31}}\right)^{10}$

USE THE POWER OF A PRODUCT/QUOTIENT

EXPRESSIONS
2m-4

Structure: Sage-N-Scribe

Simplify each expression. All exponents must be positive.

1. $\left(j^2 p^5 g\right)^{11}$

2. $\left(2m^0 n^4 p^8\right)^7$

3. $\left(3u^2 w^6\right)^5$

4. $\left(\dfrac{wx^3 y^9}{y^2 w^6 x^4}\right)^4$

5. $\left(\dfrac{3k^6 q^6}{9q^8 k^5}\right)^3$

6. $\left(\dfrac{8d^2 p^3}{4d^8 p^2}\right)^5$

1. $j^{22} p^{55} g^{11}$
2. $128 n^{28} p^{56}$
3. $243 u^{10} w^{30}$

4. $\dfrac{y^{28}}{w^{20} x^4}$

5. $\dfrac{k^3}{27 g^6}$

6. $\dfrac{32 p^5}{d^{30}}$

Cooperative Learning and Algebra 1: Becky Bride
Kagan Publishing • 1 (800) 933-2667 • www.KaganOnline.com

ACTIVITY 9 EXPLORING NEGATIVE AND ZERO EXPONENTS

1. Complete the first 4 rows of the table to the right.

2. Look at the left column of the table then describe the pattern in the exponents as you look from the top of the table to the bottom of the table.

$2^4=$	
$2^3=$	
$2^2=$	
$2^1=$	

3. Based on your observations above, continue the pattern for the left column of the table.

4. Observe the numbers in the right column of the table from the top down. Describe the pattern you see.

5. Based on your observations in problem 4, continue the pattern for the right column of the table, writing your answers in fraction form when necessary.

ACTIVITY 9

EXPLORING NEGATIVE AND ZERO EXPONENTS

$3^4=$	
$3^3=$	
$3^2=$	
$3^1=$	

6. Complete the first 4 rows of the table to the right.

7. Look at the left column of the table; then describe the pattern in the exponents as you look from the top of the table to the bottom of the table.

8. Based on your observations above, continue the pattern for the left column of the table.

9. Observe the numbers in the right column of the table from the top down. Describe the pattern you see.

10. Based on your observations in problem 9, continue the pattern for the right column of the table, writing your answers in fraction form when necessary.

Cooperative Learning and Algebra 1: Becky Bride
Kagan Publishing • 1 (800) 933-2667 • www.KaganOnline.com

ACTIVITY 9
EXPLORING NEGATIVE AND ZERO EXPONENTS

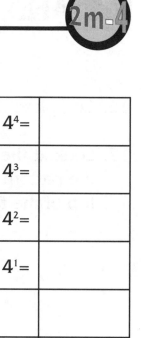

$4^4 =$	
$4^3 =$	
$4^2 =$	
$4^1 =$	

11. Complete the first 4 rows of the table to the right.

12. Look at the left column of the table; then describe the pattern in the exponents as you look from the top of the table to the bottom of the table.

13. Based on your observations above, continue the pattern for the left column of the table.

14. Observe the numbers in the right column of the table from the top down. Describe the pattern you see.

15. Based on your observations in problem 14, continue the pattern for the right column of the table, writing your answers in fraction form when necessary.

EXPLORING NEGATIVE AND ZERO EXPONENTS

ACTIVITY 9

16. Complete the first 4 rows of the table to the right.

17. Look at the left column of the table; then describe the pattern in the exponents as you look from the top of the table to the bottom of the table.

18. Based on your observations above, continue the pattern for the left column of the table.

19. Observe the numbers in the right column of the table from the top down. Describe the pattern you see.

$5^4=$	
$5^3=$	
$5^2=$	
$5^1=$	

20. Based on your observations in problem 19, continue the pattern for the right column of the table, writing your answers in fraction form when necessary.

21. Look at the row in each table that has zero as an exponent. Describe the pattern that you see.

22. Based on your description in problem 21, evaluate 62^0.

Cooperative Learning and Algebra 1: Becky Bride
Kagan Publishing • 1 (800) 933-2667 • www.KaganOnline.com

EXPLORING NEGATIVE AND ZERO EXPONENTS

EXPRESSIONS

2m-4

23. Look at the rows in each table that has −1 as an exponent. Describe the pattern that you see.

24. Based on your description in problem 23, simplify 23^{-1}.

25. Look at the rows in each table that has −2 as an exponent. Describe the pattern that you see.

26. Based on your description in problem 25, evaluate 8^{-2}.

27. Look at the rows in each table that has −3 as an exponent. Describe the pattern that you see.

28. Based on your description in problem 27, evaluate 6^{-3}.

ACTIVITY 10 REWRITE ME WITH POSITIVE EXPONENTS

Structure: Sage-N-Scribe

Rewrite each expression without negative or zero exponents.

1. w^{-4}

2. $\dfrac{1}{h^{-5}}$

3. $\dfrac{w^{-3}}{d^{0}}$

4. $k^{6}p^{-7}$

5. $\dfrac{x^{4}w^{-8}}{d^{0}g^{-7}h^{5}}$

6. $\dfrac{m^{-8}w^{-3}}{v^{-5}j^{0}h^{7}}$

Answers:

1. $\dfrac{1}{w^{4}}$

2. h^{5}

3. $\dfrac{1}{w^{3}}$

4. $\dfrac{k^{6}}{p^{7}}$

5. $\dfrac{x^{4}g^{7}}{w^{8}h^{5}}$

6. $\dfrac{v^{5}}{m^{8}w^{3}h^{7}}$

Cooperative Learning and Algebra 1: Becky Bride
Kagan Publishing • 1 (800) 933-2667 • www.KaganOnline.com

11 DISTRIBUTIVE PROPERTY REVISITED

Structure: Sage-N-Scribe

Use the distributive property to simplify.

1. $4x(3x^2 - 8x + 3)$ 2. $y^2(6 - 3y - y^2)$

3. $-m^3(6m - 8)$ 4. $8w(w^3 - 7w + 5)$

5. $2p(3p^4 - 5p + 1)$ 6. $-h^6(2h^5 - 3h^2)$

Answers:
1. $12x^3 - 32x^2 + 12x$
2. $6y^2 - 3y^3 - y^4$
3. $-6m^4 + 8m^3$
4. $8w^4 - 56w^2 + 40w$
5. $6p^5 - 10p^2 + 2p$
6. $-2h^{11} + 3h^8$

ACTIVITY

12 FACTOR ME—REVISITED

Structure: Sage-N-Scribe

Factor each expression.

1. $14x^3 - 8x^2 + 6x$

2. $28w^2 - 16w^3 + 8w^5$

3. $9h^4 + 12h^2 - 15h$

4. $-15y^7 + 20y^5 - 35y^3$

5. $3g^7 - 2g^4 + 5g^2$

6. $7x + 8xy + 9x^2y$

Answers:
1. $2x(7x^2 - 4x + 3)$
2. $4w^2(7 - 4w + 2w^3)$
3. $3h(3h^3 + 4h - 5)$
4. $-5y^3(3y^4 - 4y^2 + 7)$
5. $g^2(3g^5 - 2g^2 + 5)$
6. $x(7 + 8y + 9xy)$

Cooperative Learning and Algebra 1: Becky Bride
Kagan Publishing • 1 (800) 933-2667 • www.KaganOnline.com

EQUATIONS AND INEQUALITIES

In this chapter, students will process the meaning of a solution to an equation or inequality, how to graph equations and inequalities on a number line, and how to solve equations and inequalities. Each lesson contains activities that reinforce how to translate words into algebraic notation and vice versa, which was first introduced in the expression unit. Translating from words into algebraic equations or inequalities facilitates the creation of applications near the end of each lesson. Each lesson contains activities for students to generate graphic organizers to summarize what they learned. The final activity requires students to compare and contrast equations and inequalities.

LESSON 1 EQUATIONS

ACTIVITY 1: Find My Solution
ACTIVITY 2: Getting at the Concept of Equation Solving
ACTIVITY 3: Solve My One-Step Equation
ACTIVITY 4: Solve My Two-Step Equation
ACTIVITY 5: Solve My Multi-Step Equation
ACTIVITY 6: Solve My Advanced Multi-Step Equation
ACTIVITY 7: Translate My Words into Equations
ACTIVITY 8: Translate My Equation into Words
ACTIVITY 9: Create and Solve an Application
ACTIVITY 10: What Did We Learn?

LESSON 2 INEQUALITIES

ACTIVITY 1: Graph Me on a Number Line
ACTIVITY 2: Can You Write My Inequality?
ACTIVITY 3: Exploring Operations on Inequalities
ACTIVITY 4: Solve My One-Step Inequality
ACTIVITY 5: Solve My Two-Step Inequality
ACTIVITY 6: Solve My Multi-Step Inequality
ACTIVITY 7: Solve My Advanced Multi-Step Inequality
ACTIVITY 8: Translate My Words into an Inequality
ACTIVITY 9: Translate My Inequality into Words
ACTIVITY 10: Create and Solve an Application
ACTIVITY 11: What Did We Learn?
ACTIVITY 12: Complete My Venn Diagram

LESSON 3 PROPORTIONS

ACTIVITY 1: Write My Ratio
ACTIVITY 2: Exploring Proportions
ACTIVITY 3: Am I a Proportion?
ACTIVITY 4: Solve My Proportion
ACTIVITY 5: Scales and Proportions
ACTIVITY 6: Exploring Similar Figures
ACTIVITY 7: Proportions and Similar Figures
ACTIVITY 8: Apply Me
ACTIVITY 9: What Did We Learn?

LESSON 1
EQUATIONS

This lesson focuses on equations. The equation lesson begins with what it means to be a solution of an equation. Activity two is a teacher demonstration that gets at the concept of equation solving in a rather unorthodox way, which is awesome! Activities that progress in difficulty follow to process solving equations. Two activities are included to prepare the students for writing applications for equations. Those applications will be generated as they were for the expressions. Finally, the students will develop a graphic organizer to summarize what they have learned.

1 FIND MY SOLUTION

This first activity gives the students a replacement set and asks the students to determine which element is the solution. This reinforces the expression unit and order of operations. This activity helps students understand what they are looking for when they solve equations. This activity can be done two ways—"plug–n–chug" to find the solutions algebraically, and using a graphing calculator. This brings in the graphical perspective of equation solving, which is important. To solve the equation $2x - 3 = 8$ on the graphing calculator, you must press the y= button. This brings you to the screen where equations are entered for graphing. Into $Y_1=$ input the left side of the equation: $2x - 3$. Into $Y_2=$ input the right side of the equation: 8. Make sure your graphing window is set from -10 to 10 for the x–axis and -10 to 10 for the y–axis. Press the graph key and the two lines will intersect. The solution of the equation is the x–coordinate of the ordered pair.

▶ **Structure**
  · Sage-N-Scribe

▶ **Materials**
  · 1 Transparency 3.1.1 per pair
  · 1 sheet of paper and pencil per pair

Algebraic

Setup:
In pairs, Student A is the Sage; Student B is the Scribe. Students fold a sheet of paper in half and each writes his/her name on one half.

1. The Sage gives the Scribe step-by-step instructions on how to do problem 1.

2. The Scribe records the Sage's instructions step-by-step in writing.

3. If the Sage is correct, the Scribe praises the Sage. Otherwise, the Scribe coaches, then praises.

4. Students switch roles for the next problem.

Cooperative Learning and Algebra 1: Becky Bride
Kagan Publishing • 1 (800) 933-2667 • www.KaganOnline.com

2 GETTING AT THE CONCEPT OF EQUATION SOLVING

This activity is not a student activity but a teacher demonstration. You will use a hammer, screwdriver, nail, screw, and three boards to connect—then disconnect—the boards. Students will learn that in algebra, there are tools like hammers and screwdrivers (+, −, x, ÷) that can be used to isolate a variable (the painted board) by performing the inverse operation on both sides of an equation. By first putting the boards together and then taking them apart, the students will see that the last board connected is the first board to be disconnected. This drives home the point that the order of operations for +/− and x/÷ must be done in the reverse order when solving an equation—anything attached to the variable via addition and subtraction must be disconnected before anything attached to the variable via multiplication and division. If you want students to disconnect addition by adding the opposite of the number to both sides of an equation rather than using subtraction, then as you disconnect the boards, you can emphasize that the screwdriver must be turned in the opposite direction compared to when you put them together and the opposite end of the hammer is required to disconnect the nail. When the hammer, screwdriver, nail, screw, and boards are brought out in class, you have immediately piqued the students' interest and have their undivided attention.

Making the Manipulative
The manipulative you need can be easily made. A good length for the three boards is 9-12 inches and any width 1.5 inches or above. The depth of the boards can become an issue if you don't want the nail to protrude from the bottom board. (A depth of 1.5 inches is ideal.) Paint one of the boards any color you choose. This board will ultimately be the variable and will be the bottom board when they are all assembled. Place a nonpainted board on top of the painted board and secure these with a flat-head screw (counter sunk). A flat-head screw is crucial because another board will be placed on top of this board and needs a flat surface. Place the final nonpainted board on top of the other nonpainted board and secure these with the nail (make sure the board covers the top of the screw). Disassemble these pieces before starting the activity.

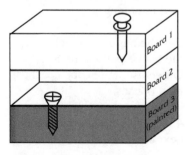

Note: A double-headed nail makes disassembling the boxes easy.

First Concept to Be Established:
Whatever you do to one side of the equation, you must do to the other side.

Notice the operative word here is "established." You will lead the students to this concept, and they will have it before you are finished with all the examples. The basic format to achieve this is

1) Write an equation on the board like 5 = 5.

2) Perform an operation to only one side of the equation, and ask if the result is still equal. Performing the operations vertically works best because the students can easily see what was done. 5 = 5 would now look like

$$\begin{array}{r} 5 = 5 \\ +3 \\ \hline 8 = 5 \end{array}$$

The students instantly see that the equation is no longer equal.

3) Ask the students what can be done to fix it so it is equal without changing the left side. The response will instantly be to add 3 to the other side.

▶ Structure
· Teacher Demonstration

▶ Materials
· 3 boards, 1-flat head screw, and 1 double-headed nail (preferably predrilled for easy use)
· 1 hammer
· 1 screwdriver

4) Repeat steps 1-3 with a new equation and a new operation each time, sometimes performing the operation on the left of the equation and sometimes performing the operation on the right side of the equation. Use subtraction, multiplication, division, squaring, and square rooting.

5) Tell the students that you are trying to make a point, and ask them what that point may be.

The Demonstration:
1. Begin with the three boards apart. Explain that you need to put the three boards together. The tools you have are a hammer and a screwdriver, and the connectors are a nail and a screw.

2. Pick up the painted board and another board and explain that you want to connect them with the screw. Pick up the hammer to connect the boards

Chapter 3: Equations and Inequalities   Lesson One

using the screw. Someone in the class should tell you that your tool is wrong. Ask them how they know it is the wrong tool. Emphasize that the connector then determined the tool that was used.

3. Begin to join the third board to the other two boards. Pick up the nail and the screwdriver. Someone will tell you that the screwdriver is the wrong tool. Ask the class why you can't use the screwdriver—that it worked well the first time. Ask again how they knew it was the wrong tool. Emphasize again that the connector determined the tool that was used.

4. Now the process begins to separate the boards. This is what you wanted to do from the beginning, but they had to be put together before you could take them apart. Emphasize that the goal is to get the painted board by itself. To do this, pick up the screwdriver. Someone will mention that it is the wrong tool. The logical question is "But I used it first when I put the boards

together. Why can't I use it first again?" You will have yet another opportunity to emphasize that the connector determines the tool. As you begin to take out the nail, use the same end of the hammer that you put the nail in with. The goal is to get someone to tell you that you need to use the opposite end. Remove the nail to disconnect the top board.

5. There are two boards still connected. Reiterate that the goal is to get the painted board by itself and to use the hammer to remove the screw. By now there should be many students telling you that the tool is wrong. This is yet another opportunity to emphasize that the connector determines which tool is required. When using the screwdriver, turn it the wrong way since that was how you used it when you connected the boards. Someone will tell you to turn it in the opposite direction and capitalize on "opposite." Remove the screw to disconnect the middle board.

6. The boards are now apart. Emphasize one more time that the goal—getting the painted board by itself—was met.

Making the Connection:
Ask the students what this demonstration has to do with math and they will quickly tell you "absolutely nothing." That is when you can tell them that it has everything to do with math. The painted board represents the variable in an equation. The goal of equation solving is to get the variable by itself on one side of the equation. In order to do that, the students must decide which numbers are connected to the variable, how they are connected, and what tool (+, −, x or ÷) they will use to disconnect. During the disconnect stage, whatever is done to one side of an equation must be done to the other side of the equation. Modeling one-step equations, emphasizing what number is connected to the variable, how it is connected, and what tool is necessary to disconnect the number from the variable, solidifies the lesson.

SOLVE MY ONE-STEP EQUATION

Note: Prior to actually solving the equations in this activity, a RallyRobin could be done using the transparency where each student tells his/her partner a) what is connected to the variable, b) what operation is connecting them, and c) what operation is needed to disconnect them. This would be helpful in a two-year algebra course. Also included on this transparency are equations involving only variables that the students will use to solve for x. Because they understand the concept, they will not skip a beat when they solve these equations.

▶ **Structure**
• Sage-N-Scribe

▶ **Materials**
• Transparency 3.1.2
• 1 sheet of paper and pencil per pair

Algebraic

Setup:
In pairs, Student A is the Sage; Student B is the Scribe. Students fold a sheet of paper in half and each writes his/her name on one half.

1. The Sage tells the Scribe how to solve problem 1.

2. The Scribe records the Sage's instructions step-by-step in writing.

3. If the Sage is correct, the Scribe praises the Sage. Otherwise, the Scribe coaches, then praises.

4. Students switch roles for the next problem.

Cooperative Learning and Algebra 1: Becky Bride
Kagan Publishing • 1 (800) 933-2667 • www.KaganOnline.com

SOLVE MY TWO-STEP EQUATION

Algebraic

Setup:
In pairs, Student A is the Sage; Student B is the Scribe. Students fold a sheet of paper in half and each writes his/her name on one half.

1. The Sage tells the Scribe how to solve problem 1.

2. The Scribe records the Sage's instructions step-by-step in writing.

3. If the Sage is correct, the Scribe praises the Sage. Otherwise, the Scribe coaches, then praises.

4. Students switch roles for the next problem.

▶ **Structure**
• Sage-N-Scribe

▶ **Materials**
• Tranparency 3.1.3
• 1 sheet of paper and pencil per pair

SOLVE MY MULTI-STEP EQUATION

Algebraic

Setup:
In pairs, Student A is the Sage; Student B is the Scribe. Students fold a sheet of paper in half and each writes his/her name on one half.

1. The Sage tells the Scribe how to solve problem 1.

2. The Scribe records the Sage's instructions step-by-step in writing.

3. If the Sage is correct, the Scribe praises the Sage. Otherwise, the Scribe coaches, then praises.

4. Students switch roles for the next problem.

▶ **Structure**
• Sage-N-Scribe

▶ **Materials**
• Transparency 3.1.4
• 1 sheet of paper and pencil per pair

SOLVE MY ADVANCED MULTI-STEP EQUATION

Algebraic

Setup:
In pairs, Student A is the Sage; Student B is the Scribe. Students fold a sheet of paper in half and each writes his/her name on one half.

1. The Sage tells the Scribe how to solve problem 1.

2. The Scribe records the Sage's instructions step-by-step in writing.

3. If the Sage is correct, the Scribe praises the Sage. Otherwise, the Scribe coaches, then praises.

4. Students switch roles for the next problem.

Note: This activity has one problem with infinite number of solutions and one with no solution. If you don't want those types of solutions, then don't have the students do problems 7 and 8.

▶ **Structure**
• Sage-N-Scribe

▶ **Materials**
• Transparency 3.1.5
• 1 sheet of paper and pencil per pair

Chapter 3: Equations and Inequalities

Lesson One

ACTIVITY 7
TRANSLATE MY WORDS INTO EQUATIONS

▶ **Structure**
• Sage-N-Scribe

▶ **Materials**
• Transparency 3.1.6
• 1 sheet of paper and pencil per pair

Algebraic

Setup:
In pairs, Student A is the Sage; Student B is the Scribe. Students fold a sheet of paper in half and each writes his/her name on one half.

1. The Sage tells the Scribe step-by-step instructions on how to write the equation using variables and numbers.

2. The Scribe records the Sage's equation in writing.

3. If the Sage is correct, the Scribe praises the Sage. Otherwise, the Scribe coaches, then praises.

4. Students switch roles for the next problem.

ACTIVITY 8
TRANSLATE MY EQUATION INTO WORDS

▶ **Structure**
• Sage-N-Scribe

▶ **Materials**
• Transparency 3.1.7
• 1 sheet of paper and pencil per pair

Algebraic

Setup:
In pairs, Student A is the Sage; Student B is the Scribe. Students fold a sheet of paper in half and each writes his/her name on one half.

1. The Sage gives the Scribe step-by-step instructions on how to write the equation in words.

2. The Scribe records the Sage's equation word-for-word in writing.

3. If the Sage is correct, the Scribe praises the Sage. Otherwise, the Scribe coaches, then praises.

4. Students switch roles for the next problem.

Cooperative Learning and Algebra 1: Becky Bride
Kagan Publishing • 1 (800) 933-2667 • www.KaganOnline.com

9 CREATE AND SOLVE AN APPLICATION

This is a fun activity where students will create, as a team effort, applications for equations. Teammates will define variables and create story problems for their variables through a series of stages. All four teammates begin their own problem so that when the process is finished, four story problems are produced by each team and each teammate has had the opportunity to contribute to each story problem.

Algebraic

1. The teacher tells students to select an object to represent a "unit."

2. Each teammate selects something to represent a "unit" for a story problem and writes it at the top of his/her paper. For example, teammate 1 may write "minutes." Teammate 2 will likely select a different "unit."

3. After the teacher signals time, or students indicate they are all done by placing their thumbs up, they pass the paper one person clockwise.

4. Each teammate reads the "unit" his/her teammate wrote and writes a related variable for that unit. For the minutes example, the next student's variable could be "m = number of minutes to groom a dog."

5. When done, students pass their papers one teammate clockwise.

6. Teammates read the unit and variable on their new sheet and write a story problem. To continue the minutes example, "Jorge has a dog grooming business. He groomed 10 dogs today and took a 30 minute lunch. If he was at work 9 hours today, how many minutes did he take to groom each dog?"

7. When done, students pass their papers one teammate clockwise.

8. Teammates read the story problem and write an equation to represent the story problem. To continue the dog grooming example, "10m + 30 = 540."

9. When done, students pass their papers one teammate clockwise.

10. Teammates check to see if the equation written on their paper is correct. If the equation is not written correctly, the teammate coaches the student who wrote it incorrectly. Then each teammate solves the equation on his/her paper.

▶ **Structure**
  · Simultaneous RoundTable

▶ **Materials**
  · 1 sheet of paper and pencil per student

11. When done, students pass their papers one teammate clockwise.

12. Teammates check the solution on their paper for accuracy. If the solution is wrong, then the student coaches the student who solved the equation. Once all the solutions are correct, the team praises each other.

Management Tip: If students use different colored pencils and write their names at the top of each paper, it is easier to hold each individual accountable.

WHAT DID WE LEARN?

▶ **Structure**
• RoundTable Consensus

▶ **Materials**
• 1 large sheet of paper per team
• different color pen or pencil for each student on the team

Synthesis via Graphic Organizer

1. Each teammate signs his/her name in the upper-right corner of the team paper with the color pen/pencil he/she is using.

2. One teammate writes "Equations" in the center of the team paper in a rectangle.

3. Teammate 1 shares with the team one core concept he/she learned in the unit.

4. The student checks for consensus.

5. The teammates show agreement or lack of agreement with thumbs up or down.

6. If there is agreement, the students celebrate and the teammate records the core concept on the graphic organizer, connecting it with a line to the main idea, "Equations." If not, teammates discuss the response until there is agreement and then they celebrate.

7. Play continues with the next student's core concept, until all core concepts are exhausted.

8. Repeat steps 3-7, with teammates adding details to each core concept and making bridges between related ideas.

Cooperative Learning and Algebra 1: Becky Bride
Kagan Publishing • 1 (800) 933-2667 • www.KaganOnline.com

ACTIVITY
1 FIND MY SOLUTION

Structure: Sage-N-Scribe

Which of the following are solutions of the equations or inequalities below?
{–2, –1, 0, 1, 2, 3}

1. $3x - 5 = (-8)$ 2. $-2 = 4 - 2x$

3. $8m + 3 = (-13)$ 4. $7 = 3y + 7$

5. $15 + 2p > 17$ 6. $2x + 5 \geq 7$

7. $-3g + 5 < 5$ 8. $(1/2)w + 5 \leq 6$

Answers:
1. {–1}
2. {3}
3. {–2}
4. {0}
5. {2, 3}
6. {1, 2, 3}
7. {1, 1, 3}
8. {–2, –1, 0, 1, 2}

SOLVE MY ONE-STEP EQUATION

Structure: Sage-N-Scribe

Solve each equation showing your work.

1. m − (−5) = 12   2. $\dfrac{u}{4}$ = (−3)

3. −4 + p = (−11)   4. 15 = (−3w)

5. c = x + h, solve for x

6. gh = w, solve for h

7. a − b = k, solve for a

8. $\dfrac{r}{k}$ = w, solve for r

Answers:
1. m = 7
2. u = −12
3. p = −7
4. w = − 5
5. x = c − h
6. h = $\dfrac{w}{y}$
7. a = k + b
8. r = wk

Cooperative Learning and Algebra 1: Becky Bride
Kagan Publishing • 1 (800) 933-2667 • www.KaganOnline.com

ACTIVITY

4 SOLVE MY TWO-STEP EQUATION

Structure: Sage-N-Scribe

Solve each equation showing your work.

1. $5x - 3 = (-13)$

2. $4 = \dfrac{1}{2}m + 3$

3. $9 - 3p = 8$

4. $19 = (-2x) + 4$

5. $-15 = 3 - \dfrac{3}{4}w$

6. $\dfrac{x}{3} + 4 = 3$

7. $8 - \dfrac{2y}{5} = 10$

8. $7 = (-3) - 2w$

Answers:
1. $x = -2$
2. $m = 2$

3. $p = \dfrac{1}{3}$

4. $x = \dfrac{-15}{2}$

5. $w = 24$
6. $x = -3$
7. $y = -5$
8. $w = -5$

ACTIVITY
5 **SOLVE MY MULTI-STEP EQUATION**

Structure: Sage-N-Scribe

Solve each equation showing your work.

1. 4x − 3 = 2x + 17

2. 8 − 3m = 2m

3. 9 + 8p = 17 − 3p

4. 3w − (−4) = 8 − w

5. 17 + 5k = 12k

6. 2y − 7 = 10 − 7y

Answers:
1. x = 10
2. m = $\frac{8}{5}$
3. p = $\frac{8}{11}$
4. w = 1
5. k = $\frac{17}{7}$
6. y = $\frac{17}{9}$

Cooperative Learning and Algebra 1: Becky Bride
Kagan Publishing • 1 (800) 933-2667 • www.KaganOnline.com

SOLVE MY ADVANCED MULTI-STEP EQUATION

Structure: Sage-N-Scribe

Solve each problem showing your work.

1. $3(2x - 7) = 5x + 4$

2. $8m - 3 - 3m = 4(3m + 1)$

3. $-p - 2 = 9(2 - 3p) + 6$

4. $3 - 2w - 5 = 5w - 22 + 3w$

5. $3(4g + 1) - g = 8 - (3g + 2)$

6. $4 - (3h + 2) = 1 - (5 - 7h)$

7. $2(6u - 1) - 3u = 9u + 8$

8. $3(8y + 2) = 10y + 2(7y + 3)$

Answers:
1. $x = 25$
2. $m = -1$
3. $p = 1$
4. $w = 2$
5. $g = \dfrac{3}{14}$
6. $h = \dfrac{3}{5}$
7. no solution
8. all real numbers

TRANSLATE MY WORDS INTO EQUATIONS

Structure: Sage-N-Scribe

Translate each of the following into equations.

1. Three more than twice a number is forty-five.

2. Seven less than four times a number is twenty-four.

3. Seventeen is four times the sum of a number and two.

4. Thirty-one is three more than twice a number divided by three.

5. The product of five and a number is equal to seven more than three times the number.

6. Two times the difference of a number and eight is equal to nine times the number increased by four.

7. The quotient of a number and six decreased by ten is four more than the number.

8. Nine times the difference of a number and one is equal to seven less than two times the number.

Answers:
1. $3 + 2n = 45$
2. $4n - 7 = 24$
3. $17 = 4(n + 2)$
4. $31 = 3 + \dfrac{2n}{3}$
5. $5n = 7 + 3n$
6. $2(n - 8) = 9n + 4$
7. $\dfrac{n}{6} - 10 = 4 + n$
8. $9(n - 1) = 2n - 7$

Cooperative Learning and Algebra 1: Becky Bride
Kagan Publishing • 1 (800) 933-2667 • www.KaganOnline.com

ACTIVITY 8 TRANSLATE MY EQUATION INTO WORDS

Structure: Sage-N-Scribe

Translate each of the following into words.

1. $4x - 5 = 16$

2. $\dfrac{w}{3} + 7 = 15$

3. $27 = 3(r + 1)$

4. $19 = 4(2y - 5) + 3$

5. $4m - 3 = \dfrac{2m}{7} + 1$

6. $3(8p - 7) = 3p + 2$

Answers:
1. Four times a number decreased by five is sixteen.
2. Some number divided by three increased by seven is fifteen.
3. Twenty-seven is equal to three times the sum of some number and one.
4. Nineteen is equal to four times the difference of twice a number and five increased by three.
5. Four times a number decreased by three is twice the same number divided by seven increased by one.
6. Three times the difference of eight times a number and seven is equal to three times the same number increased by two.

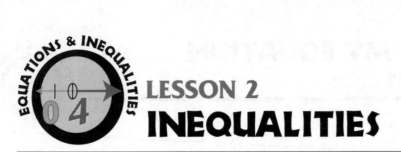

LESSON 2
INEQUALITIES

This lesson focuses on inequalities and begins with two activities on graphing inequalities on number lines. The exploratory activity leads students to discover which operations require flipping the inequality symbol when solving inequalities. In the next series of activities, students practice solving inequalities. Similar to the equation lesson, two activities are included to prepare the students to write team-generated applications that use inequalities. The students will generate a team graphic organizer to review what they have learned. The last activity requires students to generate a Venn Diagram to compare and contrast equations and inequalities.

1 GRAPH ME ON A NUMBER LINE

▶ **Structure**
• RallyCoach

▶ **Materials**
• Transparency 3.2.1
• 1 sheet of paper and pencil per pair

Connection: Algebraic to Graphical

1. Teacher poses multiple problems with Transparency 3.2.1.

2. Partner A graphs the first inequality.

3. Partner B watches and listens, checks, and praises.

4. Partner B graphs the next inequality.

5. Partner A watches and listens, checks, and praises.

6. Repeat with the remaining problems starting at step 2.

2 CAN YOU WRITE MY INEQUALITY?

▶ **Structure**
• RallyCoach

▶ **Materials**
• Transparency 3.2.2
• 1 sheet of paper and pencil per pair

Connection: Algebraic to Graphical

1. Teacher poses multiple problems with Transparency 3.2.2.

2. Partner A writes the inequality represented in the graph of problem 1.

3. Partner B watches and listens, checks, and praises.

4. Partner B writes the inequality represented in the graph of the next problem.

5. Partner A watches and listens, checks, and praises.

6. Repeat with the remaining problems starting at step 2.

ACTIVITY 3

EXPLORING OPERATIONS ON INEQUALITIES

Exploratory

Solo

1. Have each student complete the investigation individually.

Pair Consensus

2. For each problem on the investigation, each student shares with his/her partner, using RallyRobin, his/her response. They discuss the problems that they disagree on trying to come to consensus on the correct response. For the problems they can't

reach consensus on, they mark these so they can focus on them during the team phase. Encourage the students to add to their responses if their partner verbalized an understanding they did not see.

Team Consensus

3. Each pair shares their responses, using RallyRobin, with the other pair in their team, augmenting their responses if necessary. When the teams are through sharing, each student should have a detailed, complete summary.

▶ **Structure**
· Solo-Pair Consensus-Team Consensus

▶ **Materials**
· 1 Blackline 3.2.3 per student
· 1 sheet of paper and pen/pencil per student

ACTIVITY 4

SOLVE MY ONE-STEP INEQUALITY

Algebraic

Setup:
In pairs, Student A is the Sage; Student B is the Scribe. Students fold a sheet of paper in half and each writes his/her name on one half.

1. The Sage tells the Scribe how to solve problem 1.

2. The Scribe records the Sage's instructions step-by-step in writing.

3. If the Sage is correct, the Scribe praises the Sage. Otherwise, the Scribe coaches, then praises.

4. Students switch roles for the next problem.

▶ **Structure**
· Sage-N-Scribe

▶ **Materials**
· Transparency 3.2.4
· 1 sheet of paper and pencil per pair

ACTIVITY

5

SOLVE MY TWO-STEP INEQUALITY

▶ **Structure**
• Sage-N-Scribe

▶ **Materials**
• Transparency 3.2.5
• 1 sheet of paper and pencil per pair

Algebraic

Setup:
In pairs, Student A is the Sage; Student B is the Scribe. Students fold a sheet of paper in half and each writes his/her name on one half.

1. The Sage tells the Scribe how to solve problem 1.

2. The Scribe records the Sage's solution step-by-step in writing.

3. If the Sage is correct, the Scribe praises the Sage. Otherwise, the Scribe coaches, then praises.

4. Students switch roles for the next problem.

ACTIVITY

6

SOLVE MY MULTI-STEP INEQUALITY

▶ **Structure**
• Simultaneous RoundTable

▶ **Materials**
• Transparency 3.2.6
• 1 sheet of paper and pencil per pair

Algebraic

Setup:
• Teammate 1 writes problem 1 at the top of his/her paper.
• Teammate 2 writes problem 2 at the top of his/her paper.
• Teammate 3 writes problem 3 at the top of his/her paper.
• Teammate 4 writes problem 4 at the top of his/her paper.

1. All four students respond simultaneously, performing one step to solve the inequality.

2. Teacher signals time, or students place thumbs up when done with the step.

3. Students pass papers one person clockwise.

4. Students check the work of the previous student, coaching if necessary. Then the students respond simultaneously and do one additional step to solve the inequality.

5. Continue, starting at step 2 until all the inequalities are solved and checked.

7 SOLVE MY ADVANCED MULTI-STEP INEQUALITY

Algebraic

Setup:
In pairs, Student A is the Sage; Student B is the Scribe. Students fold a sheet of paper in half and each writes his/her name on one half.

1. The Sage tells the Scribe how to solve problem 1.

2. The Scribe records the Sage's instructions step-by-step in writing.

3. If the Sage is correct, the Scribe praises the Sage. Otherwise, the Scribe coaches, then praises.

4. Students switch roles for the next problem.

▶ **Structure**
· Sage-N-Scribe

▶ **Materials**
· Transparency 3.2.7
· 1 sheet of paper and pencil per pair

8 TRANSLATE MY WORDS INTO AN INEQUALITY

Algebraic

Setup:
In pairs, Student A is the Sage; Student B is the Scribe. Students fold a sheet of paper in half and each writes his/her name on one half.

1. The Sage tells the Scribe how to write the inequality in problem 1 using variables and numbers.

2. The Scribe records the Sage's instructions step-by-step in writing.

3. If the Sage is correct, the Scribe praises the Sage. Otherwise, the Scribe coaches, then praises.

4. Students switch roles for the next problem.

▶ **Structure**
· Sage-N-Scribe

▶ **Materials**
· Transparency 3.2.8
· 1 sheet of paper and pencil per pair

Chapter 3: Equations and Inequalities

Lesson Two

ACTIVITY 9
TRANSLATE MY INEQUALITY INTO WORDS

▶ **Structure**
• Sage-N-Scribe

▶ **Materials**
• Transparency 3.2.9
• 1 sheet of paper and pencil per pair

Algebraic

Setup:
In pairs, Student A is the Sage; Student B is the Scribe. Students fold a sheet of paper in half and each writes his/her name on one half.

1. The Sage tells the Scribe how to write problem 1 using words.

2. The Scribe records the Sage's instructions word for word in writing.

3. If the Sage is correct, the Scribe praises the Sage. Otherwise, the Scribe coaches, then praises.

4. Students switch roles for the next problem.

Cooperative Learning and Algebra 1: Becky Bride
Kagan Publishing • 1 (800) 933-2667 • www.KaganOnline.com

10 CREATE AND SOLVE AN APPLICATION

This is a fun activity where students will create, as a team effort, applications for inequalities. Teammates will define variables and create story problems for their variables through a series of stages. All four teammates begin their own problem so that when the process is finished, four story problems have been produced by each team and each teammate has had the opportunity to contribute to each story problem.

Algebraic

1. The teacher tells students to select an object to represent a "unit."

2. Each teammate selects something to represent a "unit" for a story problem and writes it at the top of his/her paper. For example, teammate 1 may write "dollars." Teammate 2 will likely select a different "unit."

3. After the teacher signals time, or students indicate they are all done by placing their thumbs up, they pass the paper one person clockwise.

4. Each teammate reads the "unit" his/her teammate wrote and writes a related variable. For the "dollar" example, the next student's variable could be "d = cost of a CD in dollars."

5. When done, students pass their papers one teammate clockwise.

6. Teammates read the unit and variable on their new sheet and write a story problem. To continue the "dollar" example, "Francisco purchased five CDs and paid ninety-eight cents in sales tax. His total bill was less than seventy-five dollars."

7. When done, students pass their papers one teammate clockwise.

8. Teammates read the story problem and write an inequality to represent the story problem. To continue the "dollar" example, "5d + 0.98 < 75."

9. When done, students pass their papers one teammate clockwise.

10. Teammates check to see if the inequality written on their paper is correct. If the inequality is not written correctly, the teammate coaches the student who wrote it incorrectly. Each teammate solves the inequality on his/her paper.

▶ **Structure**
· Simultaneous RoundTable

▶ **Materials**
· 1 sheet of paper and pencil per pair

11. When done, students pass their papers one teammate clockwise.

12. Teammates check the solution on their paper for accuracy. If the solution is wrong, then the student coaches the student solved the equation. Once all the solutions are correct, the team praises each other.

Management Tip: If students use different colored pencils and write their names at the top of each paper, it is easier to hold each individual accountable.

Cooperative Learning and Algebra 1: Becky Bride
Kagan Publishing • 1 (800) 933-2667 • www.KaganOnline.com

ACTIVITY

11 WHAT DID WE LEARN?

▶ **Structure**
· RoundTable Consensus

▶ **Materials**
· 1 large sheet of paper per team
· different color pen or pencil for each student on the team

Synthesis via Graphic Organizer

1. Each teammate signs his/her name in the upper-right corner of the team paper with the color pen/pencil he/she is using.

2. One teammate writes "Inequalities" in the center of the team paper in a rectangle.

3. Teammate 1 shares with the team one core concept he/she learned in the unit.

4. The student checks for consensus.

5. The teammates show agreement or lack of agreement with thumbs up or down.

6. If there is agreement, the students celebrate and the teammate records the

core concept on the graphic organizer, connecting it with a line to the main idea, "Inequalities." If not, teammates discuss the response until there is agreement and then they celebrate.

7. Play continues with the next student's core concept until all core concepts are exhausted.

8. Repeat steps 3-7, with teammates adding details to each core concept and making bridges between related ideas

ACTIVITY

12 COMPLETE MY VENN DIAGRAM

▶ **Structure**
· RoundTable Consensus

▶ **Materials**
· 1 sheet of paper per team
· 1 pencil per team

Analysis

Setup:
On the paper, a teammate will draw a Venn Diagram, with two overlapping circles or ovals, that covers the majority of the paper, similar to the one below.

1. One student tells the team something special about equations and whether this feature is unique to equations or is also true for inequalities.

2. The student checks for consensus.

3. The teammates show agreement or lack of agreement with thumbs up or down.

4. If there is agreement, the students celebrate and the student records his/her observation in the correct part of the Venn Diagram and the next student makes a response.

If not, teammates discuss the response until there is agreement, the agreement is recorded, and then they celebrate. If no agreement is reached, the steps 1-4 are repeated until the team reaches consensus.

5. Play continues with the next student's observation, repeating the entire process.

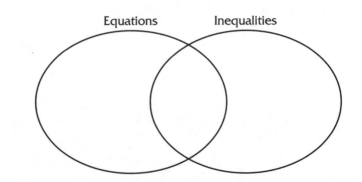

Equations   Inequalities

Cooperative Learning and Algebra 1: Becky Bride
Kagan Publishing • 1 (800) 933-2667 • www.KaganOnline.com

ACTIVITY
1 GRAPH ME ON A NUMBER LINE

Structure: RallyCoach

Graph each inequality on a number line.

1. x > 4

2. y ≤ (–3)

3. –4 ≥ m

4. –12 < p

5. 7 < w

6. 8 > n

7. v ≤ 4.5

8. h ≥ (–0.5)

Answers:

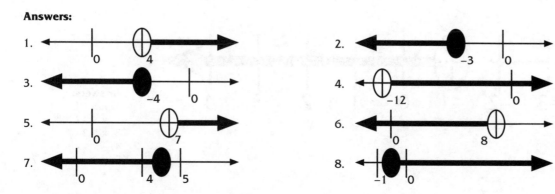

ACTIVITY
2 **CAN YOU WRITE MY INEQUALITY?**

Structure: RallyCoach

Write an inequality that fits each graph below.

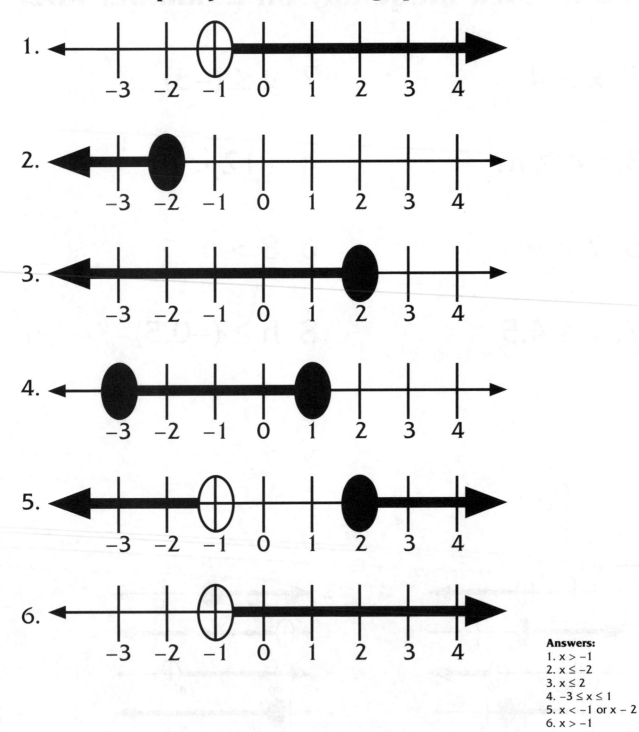

Answers:
1. x > −1
2. x ≤ −2
3. x ≤ 2
4. −3 ≤ x ≤ 1
5. x < −1 or x − 2
6. x > −1

Cooperative Learning and Algebra 1: Becky Bride
Kagan Publishing • 1 (800) 933-2667 • www.KaganOnline.com

EXPLORING OPERATIONS ON INEQUALITIES

1. a) Is the inequality $5 > 2$ a true statement?

    b) Add 4 to both sides of the inequality in part a, showing your work below.

    c) Is the inequality in part b a true statement? If not, what must be done with the inequality sign to make the statement true?

    d) Add -8 to both sides of the inequality in part a, showing your work below.

    e) Is the inequality in part d a true statement? If not, what must be done with the inequality sign to make the statement true?

2. a) Is the inequality $-8 < 1$ a true statement?

    b) Add 2 to both sides of the inequality in part a, showing your work below.

    c) Is the inequality in part b a true statement? If not, what must be done with the inequality sign to make the statement true?

    d) Add -5 to both sides of the inequality in part a, showing your work below.

EXPLORING OPERATIONS ON INEQUALITIES

e) Is the inequality in part d a true statement? If not, what must be done with the inequality sign to make the statement true?

3. Whenever you add a number to both sides of the inequality, is the resulting inequality still a true statement? **EXPLAIN.**

4. a) Is the inequality $-3 < 9$ a true statement?

   b) Subtract 8 from both sides of the inequality in part a, showing your work below.

   c) Is the inequality in part b a true statement? If not, what must be done with the inequality sign to make the statement true?

   d) Subtract -6 from both sides of the inequality in part a, showing your work below.

   e) Is the inequality in part d a true statement? If not, what must be done with the inequality sign to make the statement true?

5. a) Is the inequality $4 < -1$ a true statement?

Cooperative Learning and Algebra 1: Becky Bride
Kagan Publishing • 1 (800) 933-2667 • www.KaganOnline.com

EXPLORING OPERATIONS ON INEQUALITIES

b) Subtract 4 from both sides of the inequality in part a, showing your work below.

c) Is the inequality in part b a true statement? If not, what must be done with the inequality sign to make the statement true?

d) Subtract −1 from both sides of the inequality in part a, showing your work below.

e) Is the inequality in part d a true statement? If not, what must be done with the inequality sign to make the statement true?

6. Whenever you subtract a number from both sides of the inequality, is the resulting inequality still a true statement? **EXPLAIN.**

7. a) Is the inequality $7 > -9$ a true statement?

b) Multiply both sides of the inequality in part a by 5, showing your work below.

c) Is the inequality in part b a true statement? If not, what must be done with the inequality sign to make the statement true?

EXPLORING OPERATIONS ON INEQUALITIES

d) Multiply both sides of the inequality in part a by –7, showing your work below.

e) Is the inequality in part d a true statement? If not, what must be done with the inequality sign to make the statement true?

8. a) Is the inequality –3 < 10 a true statement?

b) Multiply both sides of the inequality in part a by 2, showing your work below.

c) Is the inequality in part b a true statement? If not, what must be done with the inequality sign to make the statement true?

d) Multiply both sides of the inequality in part a by –11, showing your work below.

e) Is the inequality in part d a true statement? If not, what must be done with the inequality sign to make the statement true?

Cooperative Learning and Algebra 1: Becky Bride
Kagan Publishing • 1 (800) 933-2667 • www.KaganOnline.com

EXPLORING OPERATIONS ON INEQUALITIES

9. Whenever you multiply both sides of an inequality by the same number, is the resulting inequality still a true statement? **EXPLAIN.**

10. a) Is the inequality 12 < 18 a true statement?

    b) Divide both sides of the inequality in part a by 3, showing your work below.

    c) Is the inequality in part b a true statement? If not, what must be done with the inequality sign to make the statement true?

    d) Divide both sides of the inequality in part a by –6, showing your work below.

    e) Is the inequality in part d a true statement? If not, what must be done with the inequality sign to make the statement true?

11. a) Is the inequality 8 > –12 a true statement?

    b) Divide both sides of the inequality in part a by 4, showing your work below.

c) Is the inequality in part b a true statement? If not, what must be done with the inequality sign to make the statement true?

d) Divide both sides of the inequality in part a by −2, showing your work below.

e) Is the inequality in part d a true statement? If not, what must be done with the inequality sign to make the statement true?

12. Whenever you divide both sides of an inequality by the same number, is the resulting inequality still a true statement? **EXPLAIN.**

13. Summarize this investigation, addressing all four operations: addition, subtraction, multiplication, and division.

Cooperative Learning and Algebra 1: Becky Bride
Kagan Publishing • 1 (800) 933-2667 • www.KaganOnline.com

SOLVE MY ONE-STEP INEQUALITY

Structure: Sage-N-Scribe

a) Solve each inequality showing your work.
b) Graph the inequality on a number line.

1. $m - (-5) > 12$

2. $\dfrac{n}{-4} \le (-3)$

3. $-5 + p \le (-11)$

4. $15 \ge (-3w)$

5. $-5 > (-2x)$

6. $x + 3 > (-3)$

7. $11 \le \dfrac{u}{-2}$

8. $-12 \ge y + 8$

Part a Answers:
1. $m > 7$
2. $n \ge 12$
3. $p \le -6$
4. $-5 \le w$
5. $x > 2.5$
6. $x > -6$
7. $u \le -22$
8. $-20 \ge y$

Cooperative Learning and Algebra 1: Becky Bride
Kagan Publishing • 1 (800) 933-2667 • www.KaganOnline.com

ACTIVITY

5 SOLVE MY TWO-STEP INEQUALITY

Structure: Sage-N-Scribe

a) Solve each inequality showing your work.
b) Graph the solution on a number line.

1. $5x - 3 < (-12)$

2. $4 > \dfrac{-1}{2}m + 5$

3. $9 - 3p \leq 8$

4. $19 \geq (-2x) + 4$

5. $-16 > 2 - \dfrac{3}{4}w$

6. $\dfrac{x}{3} - (-4) < 3$

Part a Answers:

1. $x < \dfrac{-9}{5}$

2. $2 < m$

3. $p \geq \dfrac{1}{3}$

4. $\dfrac{-15}{2} \leq x$

5. $24 < w$

6. $x < -3$

Cooperative Learning and Algebra 1: Becky Bride
Kagan Publishing • 1 (800) 933-2667 • www.KaganOnline.com

ACTIVITY 6
SOLVE MY MULTI-STEP INEQUALITY

Structure: Simultaneous RoundTable

1. $4x - 3 \geq 2x - 17$

2. $10 - 3m \leq 2m$

3. $9 - 8p < 24 - 3p$

4. $3w - (-4) > 12 - w$

Answers:
1. $-7 \leq x$
2. $2 \leq m$
3. $p > -3$
4. $w > 2$

Cooperative Learning and Algebra 1: Becky Bride
Kagan Publishing • 1 (800) 933-2667 • www.KaganOnline.com

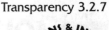

ACTIVITY 7
SOLVE MY ADVANCED MULTI-STEP INEQUALITY

Structure: Sage-N-Scribe

a) Solve ech inequality showing your work.
b) Graph the solution on a number line.

1. $-3(2x - 7) > (-5x) - 4$

2. $8m + 11 - 3m \leq 4(3m + 1)$

3. $6p + 20 \geq 2(10 + 3p) - p$

4. $3 - 3w + 5 > -5w - 7 + 3w$

5. $3(4g - 1) - 2g \leq 11(g + 1)$

6. $3 - 2(3h + 2) \leq 4 - (2h + 1)$

Part a Answers:
1. $x < 25$
2. $m \geq 1$
3. $p \geq 0$
4. $w < 15$
5. $g \geq -14$
6. $h \geq -1$

Cooperative Learning and Algebra 1: Becky Bride
Kagan Publishing • 1 (800) 933-2667 • www.KaganOnline.com

TRANSLATE MY WORDS INTO AN INEQUALITY

Structure: Sage-N-Scribe

Write each of the following as an inequality.

1. Twice a number increased by four is greater than six.

2. Four times the sum of three times a number and seven is less than or equal to ten.

3. Eight times the difference of a number and nine is greater than or equal to fifteen.

4. Five times a number decreased by four is greater than eleven increased by twice the number.

5. Lynn bought x pencils at twenty-five cents each and seven pens, which cost one dollar and fifty cents each. The total cost of her purchase was less than ten dollars.

6. Five pizzas, each with ten slices, were shared equally with an unknown number of people. The number of slices of pizza per person was less than or equal to three slices.

Answers:
1. $2n + 4 > 6$
2. $4(3n + 7) \leq 10$
3. $8(n - 9) \geq 15$
4. $5n - 4 > 11 + 2n$
5. $0.25x + 7(1.50) < 10$

6. $\dfrac{50}{x} \leq 3$

TRANSLATE MY INEQUALITY INTO WORDS

Structure: Sage-N-Scribe

Rewrite each of the following using words.

1. $5x - 3 > 16$

2. $7 \leq 2 - 3m$

3. $2(w - 5) \geq 8 - 3w$

4. $\dfrac{3 - h}{9} > 27$

5. $8 - \dfrac{2d}{7} < 15 + 9d$

6. $7(5y + 1) \geq 6y$

Answers:
1. Five times some number decreased by three is greater than sixteen.
2. Seven is less than or equal to two decreased by three times a number.
3. Two times the difference of some number and five is greater than or equal to eight decreased by three times the same number.
4. The difference of three and some number divided by nine is greater than twenty-seven.
5. Eight minus twice a number divided by seven is less than fifteen increased by nine times the same number.
6. Seven times the sum of five times some number and one is greater than or equal to the product of the same number and six.

Cooperative Learning and Algebra 1: Becky Bride
Kagan Publishing • 1 (800) 933-2667 • www.KaganOnline.com

LESSON 3
PROPORTIONS

This lesson begins with writing ratios. Activity 2 is an investigation where students explore the cross products of two ratios to discover that in a proportion, the cross products are equal. Activities follow requiring students to use cross products to identify if two ratios form a proportion and apply that principle to solve proportions. The remainder of the lesson involves applications of proportions from using scales to similar figures to ordinary applications. The lesson ends with an activity requiring the students to synthesize the lesson.

1 WRITE MY RATIO

Algebraic

Setup:
In pairs, Student A is the Sage; Student B is the Scribe. Students fold a sheet of paper in half and each writes his/her name on one half.

1. The Sage tells the Scribe how to write 2 ratios for problem 1.

2. The Scribe records the Sage's ratios in writing.

3. If the Sage is correct, the Scribe praises the Sage. Otherwise, the Scribe coaches, then praises.

4. Students switch roles for the next problem.

▶ **Structure**
· Sage-N-Scribe

▶ **Materials**
· Transparency 3.3.1
· 1 sheet of paper and pencil per pair

Chapter 3: Equations and Inequalities

Lesson Three

ACTIVITY
2 EXPLORING PROPORTIONS

▶ **Structure**
• Solo-Pair Consensus-Team Consensus

▶ **Materials**
• 1 Blackline 3.3.2 per student
• 1 calculator and pen/pencil per student

Exploratory

Solo

1. Have each student complete the investigation individually.

Pair Consensus

2. For each problem on the investigation, each student shares with his/her partner, using RallyRobin, his/her response. They discuss the problems that they disagree on, trying to come to consensus on the correct response. For the problems they can't

reach consensus on, they mark these so they can focus on them during the team phase. Encourage the students to add to their responses if their partner verbalized an understanding they did not see.

Team Consensus

3. Each pair shares their responses, using RallyRobin, with the other pair in their team, augmenting their responses if necessary. When the teams are through sharing, each student should have a detailed, complete summary.

ACTIVITY
3 AM I A PROPORTION?

▶ **Structure**
• Sage-N-Scribe

▶ **Materials**
• Transparency 3.3.3
• 1 sheet of paper and pen/pencil per pair

Algebraic

Setup:
In pairs, Student A is the Sage; Student B is the Scribe. Students fold a sheet of paper in half and each writes his/her name on one half.

1. The Sage tells the Scribe how to do problem 1.

2. The Scribe records the Sage's instructions step-by-step in writing.

3. If the Sage is correct, the Scribe praises the Sage. Otherwise, the Scribe coaches, then praises.

4. Students switch roles for the next problem.

SOLVE MY PROPORTION

Algebraic

Setup:

In pairs, Student A is the Sage; Student B is the Scribe. Students fold a sheet of paper in half and each writes his/her name on one half.

1. The Sage tells the Scribe how to solve problem 1.

2. The Scribe records the Sage's instructions step-by-step in writing.

3. If the Sage is correct, the Scribe praises the Sage. Otherwise, the Scribe coaches, then praises.

4. Students switch roles for the next problem.

▶ **Structure**
• Sage-N-Scribe

▶ **Materials**
• Transparency 3.3.4
• 1 sheet of paper and pen/ pencil per pair

SCALES AND PROPORTIONS

Algebraic

Setup:

In pairs, Student A is the Sage; Student B is the Scribe. Students fold a sheet of paper in half and each writes his/her name on one half.

1. The Sage tells the Scribe how to solve problem 1.

2. The Scribe records the Sage's instructions step-by-step in writing.

3. If the Sage is correct, the Scribe praises the Sage. Otherwise, the Scribe coaches, then praises.

4. Students switch roles for the next problem.

▶ **Structure**
• Sage-N-Scribe

▶ **Materials**
• Transparency 3.3.5
• 1 sheet of paper and pen/ pencil per pair

Chapter 3: Equations and Inequalities

Lesson Three

ACTIVITY
6

EXPLORING SIMILAR FIGURES

Note: Prerequisite skills for this exploration include plotting points on a coordinate plane and substituting numbers into the distance formula. Patty paper is the square piece of wax paper put between hamburger patties to keep them from sticking together. Tracing paper would also work.

▶ **Structure**
• Solo-Pair Consensus-Team Consensus

▶ **Materials**
• 1 Blackline 3.3.6 per student
• 1 calculator and pen/pencil per student
• 1 sheet of graph paper per student
• 1 sheet of patty paper per student

Exploratory

Solo

1. Have each student complete the investigation individually.

Pair Consensus

2. For each problem on the investigation, each student shares with his/her partner, using RallyRobin, his/her response. They discuss the problems that they disagree on, trying to come to consensus on the correct response. For the problems they can't reach consensus on, they mark these so they can focus on them during the team phase. Encourage the students to add to their responses if their partner verbalized an understanding they did not see.

Team Consensus

3. Each pair shares their responses, using RallyRobin, with the other pair in their team, augmenting their responses if necessary. When the teams are through sharing, each student should have a detailed, complete summary.

ACTIVITY
7

PROPORTIONS AND SIMILAR FIGURES

▶ **Structure**
• Sage-N-Scribe

▶ **Materials**
• Transparency 3.3.7
• 1 sheet of paper and pen/pencil per pair of students

Algebraic

Setup:
In pairs, Student A is the Sage; Student B is the Scribe. Students fold a sheet of paper in half and each writes his/her name on one half.

1. The Sage gives the Scribe step-by-step instructions on how to solve problem 1.

2. The Scribe records the Sage's solution step-by-step in writing.

3. If the Sage is correct, the Scribe praises the Sage. Otherwise, the Scribe coaches, then praises.

4. Students switch roles for the next problem or task

ACTIVITY

8 APPLY ME

Algebraic

Setup:
In pairs, Student A is the Sage; Student B is the Scribe. Students fold a sheet of paper in half and each writes his/her name on one half.

1. The Sage gives the Scribe step-by-step instructions on how to solve problem 1.

2. The Scribe records the Sage's solution step-by-step in writing.

3. If the Sage is correct, the Scribe praises the Sage. Otherwise, the Scribe coaches, then praises.

4. Students switch roles for the next problem or task

▶ **Structure**
· Sage-N-Scribe

▶ **Materials**
· Transparency 3.3.8
· 1 sheet of paper and pen/ pencil per pair

ACTIVITY

9 WHAT DID WE LEARN?

Synthesis via Graphic Organizer

1. Each teammate signs his/ her name in the upper right corner of the team paper with the color pen/pencil he/she is using.

2. One teammate writes "Proportions" in the center of the team paper in a rectangle.

3. Teammate 1 shares with the team one core concept he/she learned in the unit.

4. The student checks for consensus.

5. The teammates show agreement or lack of agreement with thumbs up or down.

6. If there is agreement, the students celebrate and the teammate records the core concept on the graphic organizer, connecting it with a line to the main idea, "Proportions." If not, team-mates discuss the response until there is agreement and then they celebrate.

7. Play continues with the next student's core concept, until all core concepts are exhausted.

8. Repeat steps 3-7, with teammates adding details to each core concept and making bridges between related ideas.

▶ **Structure**
· RoundTable Consensus

▶ **Materials**
· 1 large sheet of paper per team
· different color pen or pencil for each student on the team

Chapter 3: Equations and Inequalities

Lesson Three

1 WRITE MY RATIO

Structure: Sage-N-Scribe

Write each ratio in two different ways.

1. 5 to 8

2. 17:24

3. $\dfrac{52}{11}$

4. 26 to 41

Answers:
1. 5:8 and $\dfrac{5}{8}$
2. 17 to 24 and $\dfrac{17}{24}$
3. 52:11 and 52 to 11
4. 26:41 and $\dfrac{26}{41}$

Cooperative Learning and Algebra 1: Becky Bride
Kagan Publishing • 1 (800) 933-2667 • www.KaganOnline.com

EXPLORING PROPORTIONS

The following ratios form a proportion.

1. Find the cross product of the ratios $\dfrac{2}{3}, \dfrac{10}{15}$. Are the cross products equal or unequal?

2. Find the cross product of the ratios $\dfrac{8}{12}, \dfrac{10}{15}$. Are the cross products equal or unequal?

3. Find the cross product of the ratios $\dfrac{9}{12}, \dfrac{6}{8}$. Are the cross products equal or unequal?

4. Find the cross product of the ratios $\dfrac{15}{6}, \dfrac{20}{8}$. Are the cross products equal or unequal?

5. Study the answers above. Explain what is special about 2 ratios that form a proportion.

The following ratios do not form a proportion.

6. Find the cross product of the ratios $\dfrac{2}{3}, \dfrac{9}{15}$. Are the cross products equal or unequal?

7. Find the cross product of the ratios $\dfrac{8}{11}, \dfrac{10}{15}$. Are the cross products equal or unequal?

8. Find the cross product of the ratios $\dfrac{9}{12}, \dfrac{5}{8}$. Are the cross products equal or unequal?

9. Find the cross product of the ratios $\dfrac{15}{6}, \dfrac{20}{9}$. Are the cross products equal or unequal?

2 EXPLORING PROPORTIONS

10. Study the answers to problems 6-9 above. Explain what is true about 2 ratios that do not form a proportion.

11. Do the ratios $\dfrac{20}{15}$, $\dfrac{48}{36}$ form a proportion? Explain your answer.

12. Do the ratios $\dfrac{32}{20}$, $\dfrac{48}{28}$ form a proportion? Explain your answer.

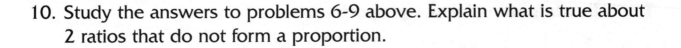

Cooperative Learning and Algebra 1: Becky Bride
Kagan Publishing • 1 (800) 933-2667 • www.KaganOnline.com

AM I A PROPORTION?

Structure: Sage-N-Scribe

Determine if the following pairs of ratios form a proportion.

1. $\dfrac{6}{9}$, $\dfrac{24}{36}$

2. $\dfrac{14}{8}$, $\dfrac{28}{18}$

3. $\dfrac{15}{12}$, $\dfrac{40}{30}$

4. $\dfrac{27}{21}$, $\dfrac{54}{42}$

Answers:
1. yes
2. no
3. no
4. yes

4 SOLVE MY PROPORTION

Structure: Sage-N-Scribe

Solve each proportion.

1. $\dfrac{20}{x} = \dfrac{8}{7}$

2. $\dfrac{9}{13} = \dfrac{p}{18}$

3. $\dfrac{10}{21} = \dfrac{8}{2w+1}$

4. $\dfrac{3h-5}{14} = \dfrac{12}{5}$

Answers:
1. 17.5
2. 12.46
3. 7.9
4. 12.87

Cooperative Learning and Algebra 1: Becky Bride
Kagan Publishing • 1 (800) 933-2667 • www.KaganOnline.com

ACTIVITY

5

SCALES AND PROPORTIONS

Structure: Sage-N-Scribe

Solve each problem below. Round answers to the hundredths place.

1. A map has a scale of 1 inch = 15 miles. If the distance between two parks is 3.5 inches on the map, what is the distance between the parks in miles?

2. An architect is making a scaled drawing of a garden that is 15.3 ft. long. If the drawing will have a scale of 1/8 inch = 1 foot, how long should the length of the garden be in his drawing?

For problems 3-4, use the description below.

An illustrator for a children's book wants to put a picture of a lion in her book. The picture is 5 inches wide and 7 inches long. To fit in the book, the picture must have a scale of 1 cm = 1.5 inches.

3. How wide is the picture in her book?

4. How long is the picture in her book?

Answers:
1. 52.5 miles
2. 1.91 in.
3. 3.33 cm
4. 4.67 cm

EXPLORING SIMILAR FIGURES

1. On a sheet of graph paper, make a coordinate plane whose x-axis scale goes from −1 to 18 and whose y-axis scale goes from −1 to 16.

2. Plot, label, and connect the following points with segments to form a geometric figure: A (0, 1), B(3, 1), C(3, 5).

3. Plot, label, and connect the following points with segments to form a geometric figure: D(6, 1), E(15, 1), F(15, 13).

4. Using the distance formula, $d = \sqrt{\left(x_2 - x_1\right)^2 + \left(y_2 - y_1\right)^2}$, find the lengths of the following segments:

   AB =   DE =

   BC =   EF =

   AC =   DF =

5. Write and simplify the ratios below:

   $\dfrac{AB}{DE} =$ $\dfrac{BC}{EF} =$ $\dfrac{AC}{DF} =$

6. What is special about the ratios above?

7. Trace ∠A onto a sheet of patty paper. Place ∠A on top of ∠D. What do you notice about these two angles?

Cooperative Learning and Algebra 1: Becky Bride
Kagan Publishing • 1 (800) 933-2667 • www.KaganOnline.com

ACTIVITY
6 EXPLORING SIMILAR FIGURES

8. Trace ∠B onto a sheet of patty paper. Place ∠B on top of ∠E. What do you notice about these two angles?

9. Trace ∠C onto a sheet of patty paper. Place ∠C on top of ∠F. What do you notice about these two angles?

10 The two triangles on your graph paper are similar. Based on your findings in problem 6 and your findings in problems 7-9, explain two properties of similar figures you have discovered.

 # PROPORTIONS AND SIMILAR FIGURES

Structure: Sage-N-Scribe

For each problem, find the value of each variable. Each pair of figures is similar. Round answers to the hundredths place.

1.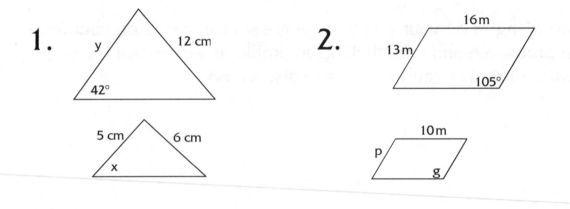

y

12 cm

42°

5 cm   6 cm

x

2.

16m

13m

105°

10m

p

g

3.

22 mm

14 mm

k

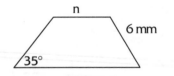

n

6 mm

35°

4.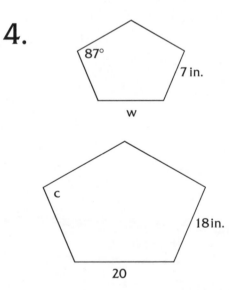

87°

7 in.

w

c

18 in.

20

Answers:
1. x = 42°; y = 10 cm
2. p = 8.13; g = 105°
3. k = 35°; n = 9.43 mm
4. c = 87°; w = 7.78 in

Cooperative Learning and Algebra 1: Becky Bride
Kagan Publishing • 1 (800) 933-2667 • www.KaganOnline.com

8 APPLY ME

Structure: Sage-N-Scribe

Solve each proportion. Round answers to the tenths place.

1. A recipe calls for 3 cups of flour for 5 dozen cookies. How much flour is needed for 8 dozen cookies?

2. A certain medication's dosage is 125 mg for every 50 lbs of body weight. How many milligrams of medication is needed for a person who weighs 180 lbs?

3. A bag of fertilizer says that 10 lbs of fertilizer will cover 35,00 sq. ft. How many square feet will 25 lbs of fertilizer cover?

4. A photo has a width of 3 inches and length of 6 inches. The photo needs to be enlarged so the width measures 5 inches. How long will the new photo be?

Answers:
1. 4.8 cups
2. 450 mg
3. 8,750 sq. ft.
4. 10 inches

LINEAR FUNCTIONS AND VERTICAL LINES

This chapter is devoted to lines. Many activities processing coordinate plane vocabulary begin the chapter. The "Exploring Intercepts" activity examines intercepts graphically so students can develop a numeric understanding of intercepts so that can be applied algebraically. An in-depth lesson on graphs of lines follows. Slope is explored and interpreted in real-world settings. Y-intercepts are presented from a second perspective. This lesson ends with graphing oblique, horizontal, and vertical lines. The third lesson is devoted to writing equations of lines and includes applications. Each lesson ends with students constructing a graphic organizer to analyze and synthesize what they have learned.

LESSON 1 COORDINATE PLANE VOCABULARY

ACTIVITY 1: Use Coordinate Plane Vocabulary
ACTIVITY 2: Exploring Intercepts
ACTIVITY 3: Find My X- and Y-Intercepts
ACTIVITY 4: Give a Set of Coordinates
ACTIVITY 5: Exploring Solutions
ACTIVITY 6: Am I a Solution? (Graphical Perspective)
ACTIVITY 7: Am I a Solution? (Algebraic Perspective)
ACTIVITY 8: What Did We Learn?

LESSON 2 GRAPHING LINES

ACTIVITY 1: Exploring Linear Graphs: Part 1
ACTIVITY 2: Name My Slope
ACTIVITY 3: Calculate My Slope, Given a Graph
ACTIVITY 4: Calculate My Slope, Given Two Points
ACTIVITY 5: Exploring the Slope of Horizontal and Vertical Lines
ACTIVITY 6: Use My Slope to Identify Me

ACTIVITY 7: Name Two Points That Form a Line with My Slope
ACTIVITY 8: Exploring Linear Graphs: Part 2
ACTIVITY 9: Name My Y-Intercept
ACTIVITY 10: Name My Slope and Y-Intercept— Advanced
ACTIVITY 11: Interpret My Slope and Intercepts
ACTIVITY 12: Exploring Horizontal and Vertical Lines
ACTIVITY 13: Graph My Line
ACTIVITY 14: What Did We Learn?

LESSON 3 WRITING EQUATIONS OF LINES

ACTIVITY 1: Write My Equation, Given a Graph
ACTIVITY 2: Write My Equation, Given a Slope and a Y-Intercept
ACTIVITY 3: Write My Equation, Given Two Points
ACTIVITY 4: Write the Equation from My Riddle
ACTIVITY 5: Linear Applications
ACTIVITY 6: Create an Application
ACTIVITY 7: What Did We Learn?

LESSON 1
COORDINATE PLANE VOCABULARY

The words "quadrant," "axis," "x-coordinate," "y-coordinate," "x-intercept," "y-intercept," and "solutions" are the focus of this lesson. In Activity 1, the students are asked to explain what "x-coordinate" means and the purpose of this is to get the students to verbalize that the x-coordinate states how far and in what direction the horizontal movement must be in order to plot the point. So often students confuse "x-coordinate" with "x-axis." The purpose of this lesson coming first is so students can use their new vocabulary on the intercept exploration. Students explore intercepts numerically and graphically in Activity 2. Activity 4 is a higher-level thinking activity, requiring students to write coordinates of points that fit a set criteria. Students will explore solutions to a function graphically and algebraically, then process what they discovered. The final activity gives students the opportunity to synthesize what they have learned.

ACTIVITY

1 USE COORDINATE PLANE VOCABULARY

▶ **Structure**
· Sage-N-Scribe

▶ **Materials**
· Transparency 4.1.1
· 1 sheet of paper and pencil per pair

Numeric and Graphical

Setup:
In pairs, Student A is the Sage; Student B is the Scribe. Students fold a sheet of paper in half and each writes his/her name on one half.

1. The Sage gives the Scribe step-by-step instruction on how to solve problem 1.

2. The Scribe records the Sage's instructions in writing.

3. If the Sage is correct, the Scribe praises the Sage. Otherwise, the Scribe coaches, then praises.

4. Students switch roles for the next problem.

Cooperative Learning and Algebra 1: Becky Bride
Kagan Publishing • 1 (800) 933-2667 • www.KaganOnline.com

2 EXPLORING INTERCEPTS

This activity begins examining six graphs and writing the coordinates of the x-intercepts. The graphs include polynomial graphs, an absolute value function graph, and a radical function graph. This exploratory exercise is laying the foundation for intercepts of any function so students don't get the wrong impression that intercepts are special for linear functions only. Intercepts will be revisited in the quadratic unit. Problems 5, 6, 11, and 12 will be a stretch for many students. This is where students are asked to generalize their numeric understanding to find intercepts algebraically. For this reason, the pair and team part of the investigation becomes even more important. The students that can make that leap will help the other students do so. The second part of the activity does for y-intercepts what the first part did for x-intercepts. The purpose of the investigation is to give students a conceptual understanding of intercepts that can be generalized to other functions and relations.

Exploratory

Solo

1. Have each student complete the investigation individually.

Pair Consensus

2. For each problem on the investigation, each student shares with his/her partner, using RallyRobin, his/her response. They discuss the problems that they disagree on, trying to come to consensus on the correct response. For the problems they can't reach consensus on, they mark these so they can focus on them during the team phase. Encourage the students to add to their responses if their partner verbalized an understanding they did not see.

Team Consensus

3. Each pair shares their responses, using RallyRobin, with the other pair in their team, augmenting their responses if necessary. When the teams are through sharing, each student should have a detailed, complete summary.

▶ **Structure**
· Solo-Pair Consensus-Team Consensus
▶
Materials
· 1 Blackline 4.1.2 per student
· 1 sheet of paper and pen/ pencil per student

ACTIVITY 3

FIND MY X- AND Y-INTERCEPTS

▶ **Structure**
• Sage-N-Scribe

▶ **Materials**
• Transparency 4.1.3
• 1 sheet of paper and pencil per pair

Algebraic

Setup:
In pairs, Student A is the Sage; Student B is the Scribe. Students fold a sheet of paper in half and each writes his/her name on one half.

1. The Sage gives the Scribe step-by-step instructions on how to find the x-intercept then the y-intercept for problem 1.

2. The Scribe records the Sage's instructions in writing.

3. If the Sage is correct, the Scribe praises the Sage. Otherwise, the Scribe coaches, then praises.

4. Students switch roles for the next problem.

ACTIVITY 4

GIVE A SET OF COORDINATES

▶ **Structure**
• RallyCoach

▶ **Materials**
• Transparency 4.1.4
• 1 sheet of paper and pencil per pair

Higher Level Thinking

1. Teacher poses problem 1 that has multiple answers using Transparency 4.1.4.

2. Partner A writes as many ordered pairs as he/she can that meet the criteria of problem 1 in 20 seconds.

3. Partner B watches and listens, coaches, and praises.

4. Partner B writes as many ordered pairs as he/she can that meet the criteria of the next problem in 20 seconds.

5. Partner A watches and listens, coaches, and praises.

6. Repeat for the remaining problems starting at step 2.

Cooperative Learning and Algebra 1: Becky Bride
Kagan Publishing • 1 (800) 933-2667 • www.KaganOnline.com

EXPLORING SOLUTIONS

Exploratory

Solo

1. Have each student complete the investigation individually.

Pair Consensus

2. For each problem on the investigation, each student shares with his/her partner, using RallyRobin, his/her response. They discuss the problems that they disagree on, trying to come to consensus on the correct response. For the problems they can't reach consensus on, they mark these so they can focus on them during the team phase. Encourage the students to add to their responses if their partner verbalized an understanding they did not see.

Team Consensus

3. Each pair shares their responses, using RallyRobin, with the other pair in their team, augmenting their responses if necessary. When the teams are through sharing, each student should have a detailed, complete summary.

▶ **Structure**
• Solo-Pair Consensus-Team Consensus

▶ **Materials**
• 1 Blackline 4.1.5 per student

AM I A SOLUTION? (GRAPHICAL PERSPECTIVE)

Graphical

1. Teacher poses multiple problems using Transparency 4.1.6.

2. In pairs, students take turns stating, orally, whether the point is a solution of the linear function graphed and explain their reasoning.

▶ **Structure**
• RallyRobin

▶ **Materials**
• Transparency 4.1.6

ACTIVITY 7
AM I A SOLUTION?
(ALGEBRAIC PERSPECTIVE)

▶ **Structure**
- Sage-N-Scribe

▶ **Materials**
- Transparency 4.1.7
- 1 sheet of paper and pencil per pair

Algebraic

Setup:
In pairs, Student A is the Sage; Student B is the Scribe. Students fold a sheet of paper in half and each writes his/her name on one half.

1. The Sage gives the Scribe step-by-step instructions on how to solve problem 1.

2. The Scribe records the Sage's instructions in writing.

3. If the Sage is correct, the Scribe praises the Sage. Otherwise, the Scribe coaches, then praises.

4. Students switch roles for the next problem.

ACTIVITY 8
WHAT DID WE LEARN?

▶ **Structure**
- RoundTable Consensus

▶ **Materials**
- 1 large sheet of paper per team
- different color pen or pencil for each student on the team

Synthesis via Graphic Organizer

1. Each teammate signs his/her name in the upper-right corner of the team paper with the color pen/pencil he/she is using.

2. One teammate writes "Coordinate Plane Vocabulary" in the center of the team paper in a rectangle.

3. Teammate 1 shares with the team one core concept he/she learned in the unit.

4. The student checks for consensus.

5. The teammates show agreement or lack of agreement with thumbs up or down.

6. If there is agreement, the students celebrate and the teammate records the core concept on the graphic organizer connecting it with a line to the main idea, "Coordinate Plane Vocabulary." If not, teammates discuss the response until there is agreement and then they celebrate.

7. Play continues with the next student's core concept until all core concepts are exhausted.

8. Repeat steps 3-7, with teammates adding details to each core concept and making bridges between related ideas.

Cooperative Learning and Algebra 1: Becky Bride
Kagan Publishing • 1 (800) 933-2667 • www.KaganOnline.com

ACTIVITY 1
USE COORDINATE PLANE VOCABULARY

Structure: Sage-N-Scribe

For each point:
a) Plot the point.
b) Identify the quadrant or axis it lies in/on.
c) Identify the x-coordinate and explain what it means.
d) Identify the y-coordinate and explain what it means.

1. (5, –3)　　　2. (–2, 4)　　　3. (–6, 0)

4. (–1, –6)　　5. (2, 7)　　　6. (0, 5)

7. (–9, –2)　　8. (4, 0)

Blackline 4.1.2

An intercept is a point(s) where a graph crosses an axis. There are two types of intercepts—x-intercepts and y-intercepts. An "x-intercept" is the point(s) where a graph crosses the x-axis. A "y-intercept" is the point(s) where a graph crosses the y-axis. This investigation explores both types of intercepts.

1. In the table below, write the coordinates of all the x-intercepts for each graph.

Graph 1	Graph 2	Graph 3	Graph 4	Graph 5	Graph 6

2. Study the x-intercepts above. Numerically, what is the same about each ordered pair? Explain why this occurs for each x-intercept.

3. If you were given a set of ordered pairs, explain how you would be able to pick out the x-intercepts.

Cooperative Learning and Algebra 1: Becky Bride
Kagan Publishing • 1 (800) 933-2667 • www.KaganOnline.com

ACTIVITY

2 **EXPLORING INTERCEPTS**

4. Using your response from problem 3, identify the x-intercepts in the following set of ordered pairs.

{(2, 3), (4, –2), (5, 0), (–8, 1), (0, 4), (–10, 2), (–2, 0), (5, 9), (–4, 12), (0, – 3)}

5. Based on your numerical knowledge of an x-intercept (review answers to problems 2 and 3), how would you find the coordinates of the x-intercept given the algebraic equation $2y + 3x = 12$?

6. Explain how the numerical understanding of an x-intercept is critical for finding the x-intercept algebraically.

7. In the table below, write the coordinates of all the y-intercepts for each graph.

Graph 1	Graph 2	Graph 3	Graph 4	Graph 5	Graph 6

8. Study the y-intercepts above. Numerically, what is the same about each ordered pair? Explain why this occurs for each y-intercept.

9. If you were given a set of ordered pairs, explain how you would be able to pick out the y-intercepts.

10. Using your response from problem 9, identify the y-intercepts in the following set of ordered pairs.

{(2, 3), (4, –2), (5, 0), (–8, 1), (0, 4), (–10, 2), (–2, 0), (5, 9), (–4, 12), (0, –3)}

11. Based on your numerical knowledge of a y-intercept (review your answers to problems 8 and 9), how would you find the coordinates of the y-intercept given the algebraic equation 2y + 3x = 12?

12. Explain how the numerical understanding of a y-intercept is critical for finding the y-intercept algebraically.

Cooperative Learning and Algebra 1: Becky Bride
Kagan Publishing • 1 (800) 933-2667 • www.KaganOnline.com

ACTIVITY

2 EXPLORING INTERCEPTS

Graph 1

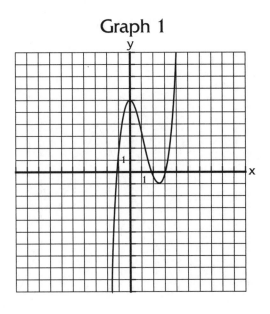

Graph 2

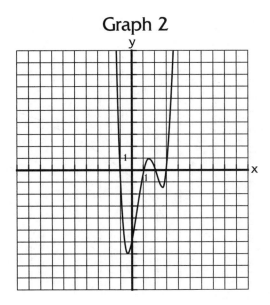

Graph 3

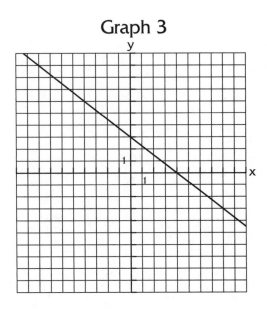

Graph 4

ACTIVITY 2 EXPLORING INTERCEPTS

Graph 5

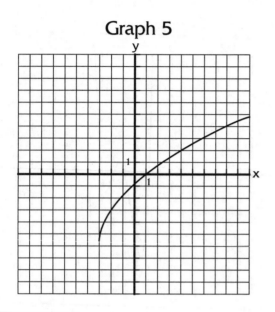

Graph 6

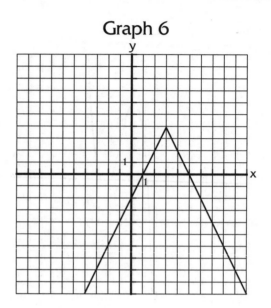

Cooperative Learning and Algebra 1: Becky Bride
Kagan Publishing • 1 (800) 933-2667 • www.KaganOnline.com

ACTIVITY

3 FIND MY X- AND Y-INTERCEPTS

Structure: Sage-N-Scribe

For each problem, find the a) x-intercept and b) y-intercept showing all work.

1. $4x - 3y = 8$

2. $9x - y = (-18)$

3. $7x + 2y = 4$

4. $5y - 3x = 9$

5. $6y = 3x - 8$

6. $3x = 7y + 3$

7. $y = 5x - 8$

8. $y = 2 - 3x$

Answers:

1. $(2, 0)$; $\left(0, \dfrac{-8}{3}\right)$

2. $(-2, 0)$; $(0, 18)$

3. $\left(\dfrac{4}{7}, 0\right)$; $(0, 2)$

4. $(-3, 0)$; $\left(0, \dfrac{9}{5}\right)$

5. $\left(\dfrac{8}{3}, 0\right)$; $\left(0, \dfrac{-4}{3}\right)$

6. $(1, 0)$; $\left(0, \dfrac{-3}{7}\right)$

7. $\left(\dfrac{8}{5}, 0\right)$; $(0, -8)$

8. $\left(\dfrac{2}{3}, 0\right)$; $(0, 2)$

GIVE A SET OF COORDINATES

Structure: RallyCoach

1. Write the coordinates of a point in quadrant II for which the x-coordinate is at least 2 units greater than the y-coordinate.

2. Write the coordinates of a point located in quadrant III for which the x-coordinate is equal to the y-coordinate.

3. Write the coordinates of a point that falls on the y-axis for which the x-coordinate is less than the y-coordinate.

4. Write the coordinates of a point in quadrant IV for which the y-coordinate is at least 2 units less than and opposite of the x-coordinate.

Cooperative Learning and Algebra 1: Becky Bride
Kagan Publishing • 1 (800) 933-2667 • www.KaganOnline.com

5 EXPLORING SOLUTIONS

Look at the two graphs below.

$$y = \frac{1}{2}x + 3 \qquad\qquad\qquad y = \frac{-2}{3}x - 3$$

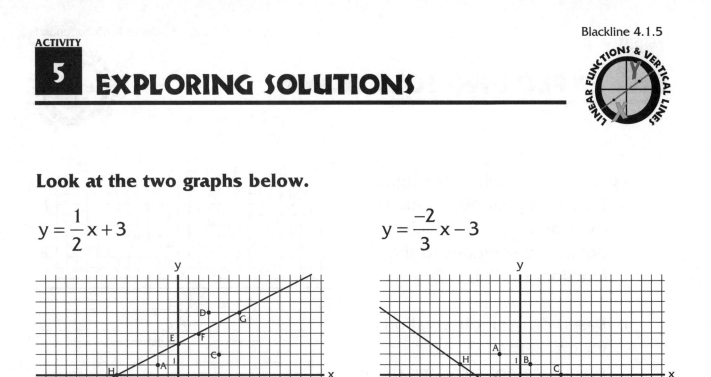

1. Points E, F, G, and H on each graph are called solutions of the function. What is special about these points and each of their graphs?

2. Points A, B, C, and D on each graph are not solutions of the function. What is the same about each of these points and each of their graphs?

3. Based on your answers to problems 1 and 2, explain what a solution to a function looks like when you look at its graph.

ACTIVITY
5

EXPLORING SOLUTIONS

4. Look at the graph to the right.
   a. Based on your observations for problems 1-3, which points are solutions to the function y = (−x) − 2?

   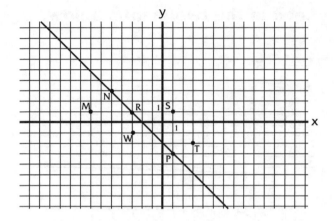

   b. Based on your observations for problems 1-3, which points are not solutions to the function y = −x − 2?

5. For the function, $y = \dfrac{1}{2}x + 3$, substitute the ordered pairs below into the function and simplify.

   a. (0, 3)   b. (2, 4)

   c. (6, 6)   d. (−6, 0)

6. Look at the simplified answers in problem 6. Describe what is special about each answer.

7. For the function, y = 2x − 3, substitute the ordered pairs below into the function and simplify.

   a. (−2, −7)   b. (−1, −5)

   c. (1, −1)   d. (2, 1)

Cooperative Learning and Algebra 1: Becky Bride
Kagan Publishing • 1 (800) 933-2667 • www.KaganOnline.com

5 EXPLORING SOLUTIONS

8. Look at the simplified answers in problem 7. Describe what is special about each answer.

9. For the function, $y = \dfrac{1}{2}x + 3$, substitute the ordered pairs below into the function and simplify.

   a. (–2, 1)   b. (–1, –3)

   c. (4, 2)   d. (3, 6)

10. Look at the simplified answers in problem 9. Describe what is special about each answer.

11. For the function, $y = 2x - 3$, substitute the ordered pairs below into the function and simplify.

    a. (–7, –2)   b. (2, –2)

    c. (4, 1)   d. (1, 6)

12. Look at the simplified answers in problem 11. Describe what is special about each answer.

EXPLORING SOLUTIONS

13. Describe how the answers in problems 6 and 8 are different from the answers in problems 10 and 12.

14. The ordered pairs in problems 5 and 7 are solutions of their respective functions and the ordered pairs in problems 9 and 11 are not solutions. Based on your answers for problems 5-13, describe how to tell algebraically whether an ordered pair is a solution of a function.

15. Based on your answer to problem 14, determine if the ordered pairs below are solutions of the function $y = (-x) - 2$. Show your work.

    a. $(-3, -1)$ b. $(1, -3)$

    c. $(1, 1)$ d. $(-3, 1)$

16. Summarize this investigation. Make sure you address how to tell if an ordered pair is a solution when given a graph or an equation.

Cooperative Learning and Algebra 1: Becky Bride
Kagan Publishing • 1 (800) 933-2667 • www.KaganOnline.com

ACTIVITY
6

AM I A SOLUTION?
(GRAPHICAL PERSPECTIVE)

Structure: RallyRobin

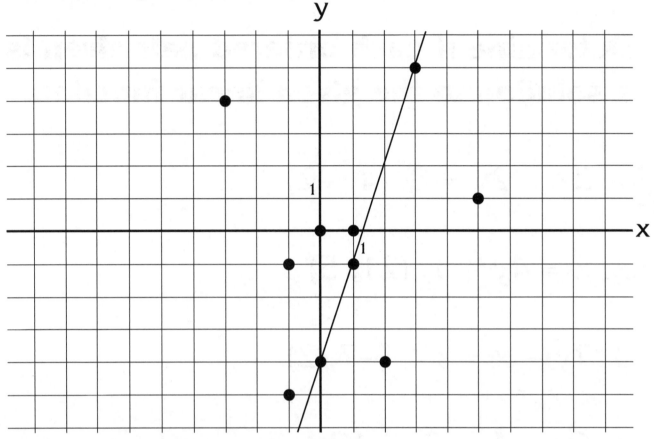

State whether the given points are solutions of the linear function above. Explain your reasoning.

1. (1, −1)   2. (3, 5)   3. (−3, 4)   4. (5, 1)

5. (2, −4)   6. (2, 2)   7. (0, −4)   8. (1, 0)

9. (−1, −5)   10. (−1, −1)

Answers:
1. yes
2. yes
3. no
4. no
5. no
6. yes
7. yes
8. no
9. no
10. no

ACTIVITY 7

AM I A SOLUTION? (ALGEBRAIC PERSPECTIVE)

Structure: Sage-N-Scribe

Determine if each ordered pair given is a solution to the given linear function.

1. $3x - 2y = 8$; $(4, -2)$

2. $x = 4y + 1$; $(21, 5)$

3. $7y - x = 12$; $(-7, 2)$

4. $6y = 4x - 2$; $(-10, -7)$

5. $3x = y + 2$; $(0, -2)$

6. $5x + 2y = 10$; $(-2, 9)$

Answers:
1. no
2. yes
3. no
4. yes
5. yes
6. no

Cooperative Learning and Algebra 1: Becky Bride
Kagan Publishing • 1 (800) 933-2667 • www.KaganOnline.com

LESSON 2
GRAPHING LINES

This lesson begins with exploring the role of "m" in the equation $y = mx + b$. The students connect the algebraic form of a linear function with its graph. Five activities follow this that process the concept of slope. One of these activities explores the slope of horizontal and vertical lines. When the students are finished with this activity, they will be able to associate undefined slope and zero slope with their graphs and numeric representations. Activity 7 explores the role of "b" in the equation $y = mx + b$ and there are two activities that follow to process what the students learned in the exploration. Activity 10 requires the students to calculate slope and the y-intercept from a graph and interpret each. The final exploratory activity investigates horizontal and vertical lines. The purpose of the activity is for students to link the numeric representation of these lines to their respective equations. Many students confuse the equations of horizontal and vertical lines and this activity gets at the core concept of why equations are written as they are and why they make sense.

1 EXPLORING LINEAR GRAPHS: PART 1

This activity is designed to be completed individually so students can discover for themselves the relationship of slope and its effect on a graph. It could begin in the pair stage if graphing calculators are in short supply. The parent graph is $y = x$. All graphs are compared to this graph until negative slope is explored. It is important to remind the students that they are comparing each new graph with the parent graph, not the graph that came before it.

Exploratory

Solo

1. Have each student complete the investigation individually.

Pair Consensus

2. For each problem on the investigation, each student shares with his/her partner, using RallyRobin, his/her response. They discuss the problems that they disagree on, trying to come to consensus on the correct response. For the problems they can't reach consensus on, they mark these so they can focus on them during the team phase. Encourage the students to add to their responses if their partner verbalized an understanding they did not see.

Team Consensus

3. Each pair shares their responses, using RallyRobin, with the other pair in their team, augmenting their responses if necessary. When the teams are through sharing, each student should have a detailed, complete summary.

▶ **Structure**
- Solo-Pair Consensus-Team Consensus

▶ **Materials**
- 1 Blackline 4.2.1 per student
- 1 pencil per student
- 1 graphing calculator per student

ACTIVITY

2

NAME MY SLOPE

This activity processes the slope part of the slope-intercept form of an equation. This activity also requires students to compare and contrast two slopes to determine the steeper slope.

▶ **Structure**
  • RallyRobin

▶ **Materials**
  • Transparency 4.2.2

Algebraic

1. Teacher poses multiple problems with Transparency 4.2.2.

2. In pairs, students take turns stating orally the slope of each function and which function has the steeper slope.

ACTIVITY

3

CALCULATE MY SLOPE, GIVEN A GRAPH

Because students have just explored slope of linear functions graphically, this activity continues the graphical theme by requiring students to calculate the slope of a line when given its graph.

▶ **Structure**
  • RallyCoach

▶ **Materials**
  • Transparency 4.2.3

Graphical

1. The teacher poses several problems using Transparency 4.2.3.

2. Partner A calculates and states the slope of the line in problem 1.

3. Partner B watches and listens, checks, and praises.

4. Partner B calculates and states the slope of the lines in the next problem.

5. Partner A watches and listens, checks, and praises.

6. Repeat for remaining problems starting at step 2.

ACTIVITY

4

CALCULATE MY SLOPE GIVEN TWO POINTS

▶ **Structure**
  • Sage-N-Scribe

▶ **Materials**
  • Transparency 4.2.4
  • 1 sheet of paper and pencil per pair

Algebraic

Setup:
In pairs, Student A is the Sage; Student B is the Scribe. Students fold a sheet of paper in half and each writes his/her name on one half.

1. The Sage gives the Scribe step-by-step instructions on how to calculate the slope in problem 1.

2. The Scribe records the Sage's instructions in writing.

3. If the Sage is correct, the Scribe praises the Sage. Otherwise, the Scribe coaches, then praises.

4. Students switch roles for the next problem.

Cooperative Learning and Algebra 1: Becky Bride
Kagan Publishing • 1 (800) 933-2667 • www.KaganOnline.com

5 EXPLORING THE SLOPE OF HORIZONTAL AND VERTICAL LINES

This activity explores undefined and zero slope. The students will link the numeric form of undefined and zero slope to the graph of the respective lines. The prerequisite skill necessary for this exploration is computing the slope of a line passing through two given points.

Exploratory

Solo

1. Have each student complete the investigation individually.

Pair Consensus

2. For each problem on the investigation, each student shares with his/her partner, using RallyRobin, his/her response. They discuss the problems that they disagree on, trying to come to consensus on the correct response. For the problems they can't reach consensus on, they mark these so they can focus on them during the team phase. Encourage the students to add to their responses if their partner verbalized an understanding they did not see.

Team Consensus

3. Each pair shares their responses, using RallyRobin, with the other pair in their team, augmenting their responses if necessary. When the teams are through sharing, each student should have a detailed, complete summary.

▶ **Structure**
• Solo-Pair Consensus-Team Consensus

▶ **Materials**
• 1 Blackline 4.2.5 per student
• 1 sheet of graph paper and pen/pencil per student

---

6 USE MY SLOPE TO IDENTIFY ME

Numeric

Setup:
In pairs, Student A is the Sage; Student B is the Scribe. Students fold a sheet of paper in half and each writes his/her name on one half.

1. The Sage gives the Scribe step-by-step instructions on how to find the slope for problem 1.

2. The Scribe records the Sage's instructions step-by-step in writing.

3. If the Sage is correct, the Scribe praises the Sage. Otherwise, the Scribe coaches, then praises.

4. Students switch roles for the next problem.

▶ **Structure**
• Sage-N-Scribe

▶ **Materials**
• Transparency 4.2.6
• 1 sheet of paper and pen/pencil per pair

ACTIVITY

7 NAME TWO POINTS THAT FORM A LINE WITH MY SLOPE

In this activity students will be given a slope and asked to name two points on a line with that slope. Each problem is one round. For problem 1, give students 1 minute to name as many pairs of points as they can. (These pairs of points do not have to be on the same line.) Students may see that these lines will be parallel and may lead them to the fact that parallel lines have equal slope.

▶ **Structure**
• RallyCoach

▶ **Materials**
• Transparency 4.2.7
• 1 sheet of graph paper and pencil per pair

Numeric

1. Teacher poses a problem with multiple answers using Transparency 4.2.7.

2. Partner A writes as many pairs of points that form lines with the given slope.

3. Partner B watches and listens, checks, and praises.

4. Partner B writes as many pairs of points that form lines with the given slope.

5. Partner A watches and listens, checks, and praises.

6. Repeat for remaining problems starting at step 2.

ACTIVITY

8 EXPLORING LINEAR GRAPHS: PART 2

This activity explores the role of "b" in the slope-intercept form of the equation of a line. The graphing window for this activity is –10 to 10 for both the x- and y-axis. Algebra 1 students probably are not experienced in adjusting the view window of a graphing calculator. This inexperience is good for questions 5 and 10 because the graphs are not in the view window, so students will have to generalize their understanding to these functions.

▶ **Structure**
• Solo-Pair Consensus-Team Consensus

▶ **Materials**
• 1 Blackline 4.2.8 per student
• 1 pencil per student
• 1 graphing calculator per student

Exploratory

Solo

1. Have each student complete the investigation individually.

Pair Consensus

2. For each problem on the investigation, each student shares with his/her partner, using RallyRobin, his/her response. They discuss the problems that they disagree on, trying to come to consensus on the correct response. For the problems they can't

reach consensus on, they mark these so they can focus on them during the team phase. Encourage the students to add to their responses if their partner verbalized an understanding they did not see.

Team Consensus

3. Each pair shares their responses, using RallyRobin, with the other pair in their team, augmenting their responses if necessary. When the teams are through sharing, each student should have a detailed, complete summary.

Cooperative Learning and Algebra 1: Becky Bride
Kagan Publishing • 1 (800) 933-2667 • www.KaganOnline.com

ACTIVITY
9   NAME MY Y-INTERCEPT

Algebraic

1. Teacher poses several problems with Transparency 4.2.2.

2. In pairs, students take turns stating, orally, the y-intercept of each function.

▶ **Structure**
· RallyRobin

▶ **Materials**
· Transparency 4.2.2

ACTIVITY
10   NAME MY SLOPE AND Y-INTERCEPT—ADVANCED

Algebraic

Setup:
In pairs, Student A is the Sage; Student B is the Scribe. Students fold a sheet of paper in half and each writes his/her name on one half.

1. The Sage gives the Scribe step-by-step instructions on how to find the slope and y-intercept for problem 1.

2. The Scribe records the Sage's instructions step-by-step in writing.

3. If the Sage is correct, the Scribe praises the Sage. Otherwise, the Scribe coaches, then praises.

4. Students switch roles for the next problem.

▶ **Structure**
· Sage-N-Scribe

▶ **Materials**
· Transparency 4.2.9
· 1 sheet of paper and pen/pencil per pair

11 INTERPRET MY SLOPE AND INTERCEPTS

This activity gives the students a graph with labels on each axis that represents a real-world application. The students are to compute the slope and interpret its meaning. The students can calculate the slope using the graph or the slope formula. If the students choose to use the graph, emphasize that they need to pay attention to the scale on each axis. Usually students who compute the slope using the slope formula, after identifying the coordinates of two points on the graph, make fewer mistakes. The students also are required to give the coordinates of the y-intercept and explain the meaning of this point.

▶ **Structure**
  · Sage-N-Scribe

▶ **Materials**
  · 1 Blackline 4.2.10 per pair
  · 1 sheet of paper and pencil per pair

Algebraic

Setup:
In pairs, Student A is the Sage; Student B is the Scribe. Students fold a sheet of paper in half and each writes his/her name on one half.

1. The Sage gives the Scribe step-by-step instructions on how to find and interpret the slope and y-intercept for problem 1.

2. The Scribe records the Sage's instructions step-by-step in writing.

3. If the Sage is correct, the Scribe praises the Sage. Otherwise, the Scribe coaches, then praises.

4. Students switch roles for the next problem.

Answers:
1. a. 0.13
   b. It costs $0.13 per ounce to ship.
   c. (0, 0)
   d. It cost $0 to ship 0 ounces.

2. a. 60
   b. The speed is 60 miles per hour.
   c. (0, 0)
   d. It has taken 0 hours to travel 0 miles.

3. a. 25
   b. The population increases 25,000 people each decade.
   c. (0, 75)
   d. In 1930 the population was 75,000 people.

4. a. 0.57
   b. It costs $.57 per mile
   c. (0, 2)
   d. It costs $2 for a taxi before the driver drives any miles.

12 EXPLORING HORIZONTAL AND VERTICAL LINES

This activity was designed to help students understand horizontal and vertical lines. They compare graphs (graphic perspective) to a table containing coordinates of points on the line (numeric perspective) to their algebraic representation (algebraic perspective). Students tend to graph lines whose equations are $y = \#$ as vertical lines because the y-axis is vertical and graph lines whose equations are $x = \#$ as horizontal lines because the x-axis is horizontal. This investigation tries to give students a deeper understanding of these lines so they will graph them correctly and be able to write their equations correctly.

Exploratory

Solo

1. Have each student complete the investigation individually.

Pair Consensus

2. For each problem on the investigation, each student shares with his/her partner, using RallyRobin, his/her response. They discuss the problems that they disagree on, trying to come to consensus on the correct response. For the problems they can't reach consensus on, they mark these so they can focus on them during the team phase. Encourage the students to add to their responses if their partner verbalized an understanding they did not see.

Team Consensus

3. Each pair shares their responses, using RallyRobin, with the other pair in their team, augmenting their responses if necessary. When the teams are through sharing, each student should have a detailed, complete summary.

> ▶ **Structure**
> • Solo-Pair Consensus-Team Consensus

> ▶ **Materials**
> • 1 Blackline 4.2.11 per student
> • 1 sheet of graph paper and pen/pencil per student

---

13 GRAPH MY LINE

This activity requires students to graph lines. The equations are given in standard form. This activity can be done twice—once generating a table of coordinates then plotting them, and the second time using the slope and y-intercept to graph them.

Algebraic to Graphical

Setup:
In pairs, Student A is the Sage; Student B is the Scribe. Students fold a sheet of paper in half and each writes his/her name on one half.

1. The Sage gives the Scribe step-by-step instructions on how to graph problem 1.

2. The Scribe graphs the line using the Sage's instructions.

3. If the Sage is correct, the Scribe praises the Sage. Otherwise, the Scribe coaches, then praises.

4. Students switch roles for the next problem.

> ▶ **Structure**
> • Sage-N-Scribe

> ▶ **Materials**
> • Transparencies 4.2.12 and 4.2.13
> • 1 sheet of graph paper and pencil per pair

Cooperative Learning and Algebra 1: Becky Bride
Kagan Publishing • 1 (800) 933-2667 • www.KaganOnline.com

ACTIVITY

14 WHAT DID WE LEARN?

▶ **Structure**
- RoundTable Consensus

▶ **Materials**
- 1 large sheet of paper per team
- different color pen or pencil for each student on the team

Synthesis via Graphic Organizer

1. Each teammate signs his/her name in the upper right corner of the team paper with the color pen/pencil he/she is using.

2. One teammate writes "Graphs of Lines" in the center of the team paper in a rectangle.

3. Teammate 1 shares with the team one core concept he/she learned in the unit.

4. The student checks for consensus.

5. The teammates show agreement or lack of agreement with thumbs up or down.

6. If there is agreement, the students celebrate and the teammate records the core concept on the graphic organizer connecting it with a line to the main idea, "Graphs of Lines." If not, teammates discuss the response until there is agreement and then they celebrate.

7. Play continues with the next student's core concept until all core concepts are exhausted.

8. Repeat steps 3-7, with teammates adding details to each core concept and making bridges between related ideas.

Cooperative Learning and Algebra 1: Becky Bride
Kagan Publishing • 1 (800) 933-2667 • www.KaganOnline.com

ACTIVITY 1 — EXPLORING LINEAR GRAPHS: PART 1

Objective: To determine the role of "m" in the function y = mx + b.

1. Graph the function y = x. This function will be called the "parent function."

    a. Describe the graph.

    b. The coefficient of "x" in the function y = mx + b is "m." With that information, what is the value of "m" in the parent function y = x?

2. Graph y = 3x. Compare this graph with the graph of the parent function. What effect did the coefficient 3 have on the graph when compared to the parent graph?

3. Graph $y = \dfrac{14}{3}x$. Compare this graph with the graph of the parent function. What effect did the coefficient $\dfrac{14}{3}$ have on the graph when compared to the parent graph?

4. Graph y = 5x. Compare this graph with the graph of the parent function. What effect did the coefficient 5 have on the graph when compared to the parent graph?

5. As "m" increased (m > 1), describe what happened to the graphs when compared with the parent graph.

6. Clear all functions from the calculator except the parent graph.

EXPLORING LINEAR
GRAPHS: PART 1

7. Graph $y = \dfrac{2}{3}x$. Compare this graph with the graph of the parent function. What effect did the coefficient $\dfrac{2}{3}$ have on the graph when compared to the parent graph?

8. Graph $y = \dfrac{3}{7}x$. Compare this graph with the graph of the parent function. What effect did the coefficient $\dfrac{3}{7}$ have on the graph when compared to the parent graph?

9. Graph $y = \dfrac{2}{11}x$. Compare this graph with the graph of the parent function. What effect did the coefficient $\dfrac{2}{11}$ have on the graph when compared to the parent graph?

10. When "m" is between 0 and 1, describe what happened to the graphs when compared to the parent graph.

11. Clear all functions from the calculator except the parent graph.

12. Graph $y = -x$.

    a. What is the value of "m" in the equation?

    b. Describe the effect the negative sign had on the graph when compared to the parent graph.

Cooperative Learning and Algebra 1: Becky Bride
Kagan Publishing • 1 (800) 933-2667 • www.KaganOnline.com

EXPLORING LINEAR GRAPHS: PART 1

13. Clear the function y = x from your calculator. The parent graph is now y = −x.

14. Graph y = −2x.. Compare this graph with the graph of the parent function. What effect did the coefficient −2 have on the graph when compared to the parent graph?

15. Graph y = $\dfrac{-10}{3}$ x. Compare this graph with the graph of the parent function. What effect did the coefficient $\dfrac{-10}{3}$ have on the graph when compared to the parent graph?

16. Graph y = − 6x. Compare this graph with the graph of the parent function. What effect did the −6 have on the graph when compared to the parent graph?

17. Describe what effect "m" has on the graphs when m < −1.

18. Clear all functions from the calculator except the parent graph.

19. Graph y = $\dfrac{-2}{3}$x. Compare this graph with the graph of the parent function. What effect did the coefficient $\dfrac{-2}{3}$ have on the graph when compared to the parent graph?

ACTIVITY 1

EXPLORING LINEAR GRAPHS: PART 1

20. Graph $y = \dfrac{-2}{5}x$. Compare this graph with the graph of the parent function. What effect did the coefficient $\dfrac{-2}{5}$ have on the graph when compared to the parent graph?

21. Graph $y = \dfrac{-1}{8}x$. Compare this graph with the graph of the parent function. What effect did the coefficient $\dfrac{-1}{8}$ have on the graph when compared to the parent graph?

22. When "m" is between -1 and 0, describe what happens to the graphs when compared to the parent graph.

23. Clear all functions including the parent graph.

24. Graph $y = 5x$ and $y = -5x$. Compare and constrast their graphs. Explain what is similar and what is different.

25. Based on your observations in this activity, complete the following.
    a. Which line will be steeper: $y = -2x$ or $y = \dfrac{21}{5}x$? Explain your reasoning.
    b. Which line will be less steep: $y = \dfrac{1}{2}x$ or $y = \dfrac{3}{2}x$? Explain your reasoning.
    c. Write the equation of a line that is less steep than $y = -7x$ and that rises as you look at the graph from left to right.

Cooperative Learning and Algebra 1: Becky Bride
Kagan Publishing • 1 (800) 933-2667 • www.KaganOnline.com

ACTIVITIES
2 & 9 NAME MY SLOPE
NAME MY Y-INTERCEPT

LINEAR FUNCTIONS & VERTICAL LINES

Structure: RallyRobin

Activity 2: For each problem, state the slope of each function and state which function has the steepest slope.
Activity 9: Give the coordinates of the y-intercept for each problem.

1. $y = 2x - 12$ and $y = 9 - 3x$

2. $y = 1 - 3x$ and $y = \dfrac{3}{4}x + 5$

3. $y = \dfrac{1}{2}x + 4$ and $y = 3 - \dfrac{5}{4}x$

4. $y = \dfrac{-1}{5}x - 1$ and $y = \dfrac{2}{21}x + 5$

Answers:

Activity 2:
1. 2; −3; line 2

2. −3; $\dfrac{3}{4}$; line 1

3. $\dfrac{1}{2}$; $\dfrac{-5}{4}$; line 2

4. $\dfrac{-1}{5}$; $\dfrac{2}{21}$; line 1

Activity 9:
1. (0, −12); (0, 9)

2. (0, 1); (0, 5)

3. (0, 4); (0, 3)

4. (0, −1); (0, 5)

CALCULATE MY SLOPE, GIVEN A GRAPH

ACTIVITY 3

Structure: RallyCoach

Calculate the slope of each line.

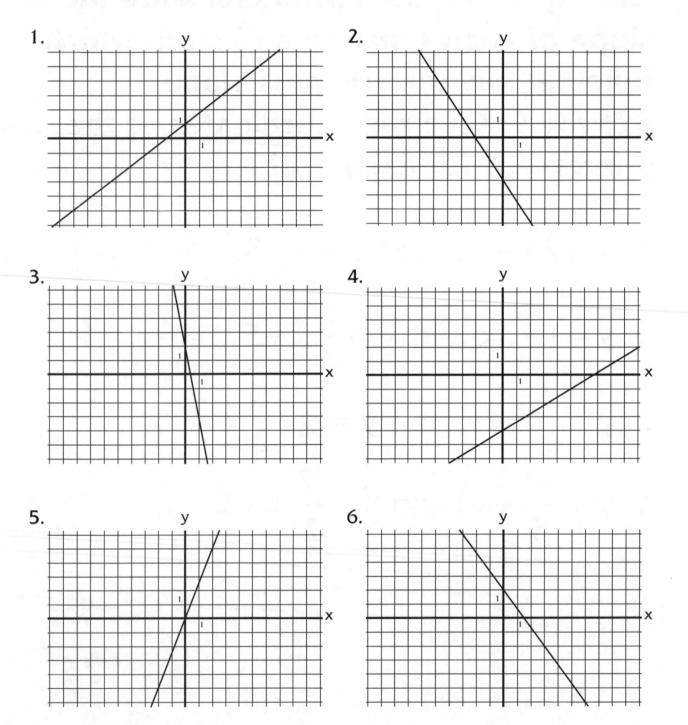

1.

2.

3.

4.

5.

6.

Cooperative Learning and Algebra 1: Becky Bride
Kagan Publishing • 1 (800) 933-2667 • www.KaganOnline.com

ACTIVITY 4 — CALCULATE MY SLOPE, GIVEN TWO POINTS

Structure: Sage-N-Scribe

Calculate the slope of the line containing the points.

1. (–3, 5) and (7, 1)

2. (–4, –3) and (–2, –7)

3. (7, 8) and (1, –5)

4. (–10, –2) and (1, 9)

5. (–3, 8) and (–6, 7)

6. (–12, 9) and (9, 4)

7. (5, –5) and (–11, 7)

8. (7, 3) and (–1, –2)

Answers:

1. $\frac{-2}{5}$

2. –2

3. $\frac{13}{6}$

4. 1

5. $\frac{1}{3}$

6. $\frac{-5}{21}$

7. $\frac{-3}{4}$

8. $\frac{5}{8}$

EXPLORING THE SLOPE OF HORIZONTAL AND VERTICAL LINES

For problems 1-4, use the ordered pairs: (–3, 5) and (2, 5).

1. Compare the ordered pairs. What is special about their coordinates? How do you think this will affect the slope of the line passing through these points?

2. Graph the points on a coordinate plane and draw a line through them. Describe the line that passes through these points.

3. Using the slope formula, calculate the slope of the line passing through these points. Leave your answer in fraction form.

4. Using your calculator, change the fraction answer in problem 3 to a decimal. What is the decimal?

For problems 5-9, use the ordered pairs: (4, –2) and (1, –2).

5. Compare the ordered pairs. What is special about their coordinates? How do you think this will affect the slope of the line passing through these points?

6. Graph the points on a coordinate plane and draw a line through them. Describe the line that passes through these points.

Cooperative Learning and Algebra 1: Becky Bride
Kagan Publishing • 1 (800) 933-2667 • www.KaganOnline.com

EXPLORING THE SLOPE OF HORIZONTAL AND VERTICAL LINES

7. Using the slope formula, calculate the slope of the line passing through these points. Leave your answer in fraction form.

8. Using your calculator, change the fraction answer in problem 7 to a decimal. What is the decimal?

9. Compare the fraction answers from problems 3 and 7. What is similar about these fractions?

For problems 10-13, use the ordered pairs: (–3, 2) and (–3, 5).

10. Compare the ordered pairs. What is special about their coordinates? How do you think this will affect the slope of the line passing through these points?

11. Graph the points on a coordinate plane and draw a line through them. Describe the line that passes through these points.

12. Using the slope formula, calculate the slope of the line passing through these points. Leave your answer in fraction form.

EXPLORING THE SLOPE OF
HORIZONTAL AND VERTICAL LINES

13. Using your calculator, change the fraction answer in problem 12 to a decimal. What is the decimal?

For problems 14-17, use the ordered pairs: (4, 7) and (4, –2).

14. Compare the ordered pairs. What is special about their coordinates? How do you think this will affect the slope of the line passing through these points?

15. Graph the points on a coordinate plane and draw a line through them. Describe the line that passes through these points.

16. Using the slope formula, calculate the slope of the line passing through these points. Leave your answer in fraction form.

17. Using your calculator, change the fraction answer in problem 16 to a decimal. What is the decimal?

18. Compare the fraction answers from problems 12 and 16. What is similar about these fractions?

Cooperative Learning and Algebra 1: Becky Bride
Kagan Publishing • 1 (800) 933-2667 • www.KaganOnline.com

EXPLORING THE SLOPE OF HORIZONTAL AND VERTICAL LINES

19. Summarize this investigation, explaining how to tell by looking at the ordered pairs when the slope of the line containing those ordered pairs is undefined and when the slope is zero. Include a description of how to tell whether a line has undefined slope or zero slope when you look at the fraction after you have calculated the slope algebraically.

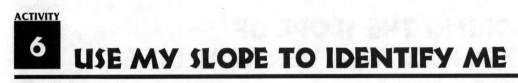

6 **USE MY SLOPE TO IDENTIFY ME**

Structure: Sage-N-Scribe

Calculate the slope of the line containing the points and tell whether the line is oblique, horizontal, or vertical based on the slope.

1. (–3, 6) and (2, 8)

2. (4, –1) and (–8, –2)

3. (7, –3) and (6, –5)

4. (4, –3) and (4, 8)

5. (–9, 7) and (–6, 7)

6. (–12, 5) and (9, 4)

7. (5, –2) and (5, 7)

8. (16, –2) and (–1, –2)

Answers:

1. $\frac{2}{5}$; oblique

2. $\frac{1}{12}$; oblique

3. +2; oblique

4. undefined; vertical

5. 0; horizontal

6. $\frac{-1}{21}$; oblique

7. undefined; vertical

8. 0; horizontal

Cooperative Learning and Algebra 1: Becky Bride
Kagan Publishing • 1 (800) 933-2667 • www.KaganOnline.com

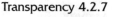

7 NAME TWO POINTS THAT FORM A LINE WITH MY SLOPE

Structure: RallyCoach

Name as many pairs of points so that the line passing through them has the given slope.

1. $\dfrac{5}{3}$

2. $\dfrac{-2}{7}$

3. undefined slope

4. 4

5. –8

6. 0

ACTIVITY 8
EXPLORING LINEAR GRAPHS: PART 2

Objective: To determine the role of "b" in the function y = mx + b.

1. Graph the function y = x. This function will be called the parent function.
    a. Describe the graph.
    b. The constant or number that is added to mx in the equation
    y = mx + b is "b." With this information, what is the value of "b" in
    the parent graph y = x? In other words, what is being added to x?
    c. What are the coordinates of the y-intercept?

2. Graph y = x + 2.
    a. What are the coordinates of the y-intercept?
    b. What is the value of "b"?
    c. What effect did adding 2 have on this graph when compared to
    the parent function?

3. Graph y = x + 5.
    a. What are the coordinates of the y-intercept?
    b. What is the value of "b"?
    c. What effect did adding 5 have on this graph when compared to
    the parent function?

4. Graph y = x + 8.
    a. What are the coordinates of the y-intercept?
    b. What is the value of "b"?
    c. What effect did adding 8 have on this graph when compared to
    the parent function?

Cooperative Learning and Algebra 1: Becky Bride
Kagan Publishing • 1 (800) 933-2667 • www.KaganOnline.com

ACTIVITY
8

EXPLORING LINEAR
GRAPHS: PART 2

5. Without graphing and based on the pattern you observed previously, answer the following questions using the function y = x + 50.
   a. What are the coordinates of the y-intercept?
   b. What is the value of "b"?
   c. What effect did adding 50 have on this graph when compared to the parent function?

6. Clear all graphs except the parent function.

7. Graph y = x − 3.
   a. What are the coordinates of the y-intercept?
   b. What is the value of "b"?
   c. What effect did subtracting 3 have on this graph when compared to the parent function?

8. Graph y = x − 6.
   a. What are the coordinates of the y-intercept?
   b. What is the value of "b"?
   c. What effect did subtracting 6 have on this graph when compared to the parent function?

9. Graph y = x − 9.
   a. What are the coordinates of the y-intercept?
   b. What is the value of "b"?
   c. What effect did subtracting 9 have on this graph when compared to the parent function?

EXPLORING LINEAR GRAPHS PART 2

10. Without graphing and based on the pattern you observed in problems 7-9, answer the following questions using the function y = x – 72.
    a. What are the coordinates of the y-intercept?
    b. What is the value of "b"?
    c. What effect did subtracting 72 have on this graph when compared to the parent function?

11. Write a paragraph summarizing what you learned from this investigation.

Cooperative Learning and Algebra 1: Becky Bride
Kagan Publishing • 1 (800) 933-2667 • www.KaganOnline.com

ACTIVITY
10 NAME MY SLOPE AND Y-INTERCEPT—ADVANCED

Structure: Sage-N-Scribe

For each problem, state the slope and y-intercept.

1. $3x - 4y = 12$

2. $5y + x = 6$

3. $2y + 4x = 10$

4. $7x - y = 11$

5. $16 = 3x - 4y$

6. $15 = 2x - 3y$

Answers:

1. $\frac{3}{4}$; $(0, -3)$

2. $\frac{-1}{5}$; $\left(0, \frac{6}{5}\right)$

3. -2, $(0, 5)$

4. 7; $(0, -11)$

5. $\frac{3}{4}$; $(0, -4)$

6. $\frac{2}{3}$; $(0, -5)$

ACTIVITY 11 INTERPRET MY SLOPE AND INTERCEPTS

Structure: Sage-N-Scribe

For each problem:

a) Calculate the slope, rounding to the hundredths place.

b) Explain what the slope means.

c) Write the coordinates of the y-intercept.

d) Explain what the y-intercept means.

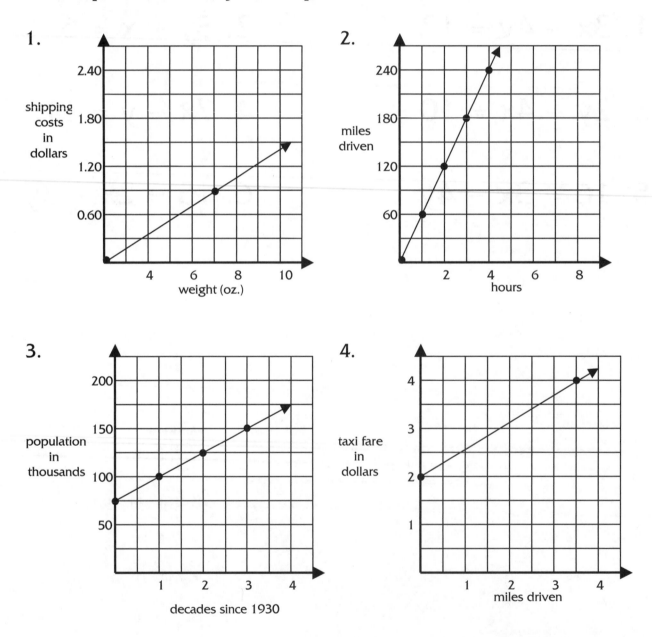

1.

shipping costs in dollars

2.40
1.80
1.20
0.60

4   6   8   10

weight (oz.)

2.

miles driven

240
180
120
60

2   4   6   8

hours

3.

population in thousands

200
150
100
50

1   2   3   4

decades since 1930

4.

taxi fare in dollars

4
3
2
1

1   2   3   4

miles driven

Answers are on teachers notes (page 246).

Cooperative Learning and Algebra 1: Becky Bride
Kagan Publishing • 1 (800) 933-2667 • www.KaganOnline.com

EXPLORING HORIZONTAL AND VERTICAL LINES

For problems 1-4, use the table to the right.

x	y
1	4
−2	4
6	4
−3	4
5	4

1. Examine the table of values at the right. What is special about this table?

2. On a coordinate plane, graph each ordered pair and draw a line through them. Describe the line that these ordered pairs form.

3. What is the slope of this line?

4. The equation of this line is y = 4. Why is the equation of this line y = 4?

For problems 5-8, use the table to the right.

x	y
2	−3
4	−3
−1	−3
−3	−3
0	−3

5. Examine the table of values at the right. What is special about this table?

6. On a coordinate plane, graph each ordered pair and draw a line through them. Describe the line that these ordered pairs form.

7. What is the slope of this line?

EXPLORING HORIZONTAL AND VERTICAL LINES

8. The equation of this line is $y = -3$. Why is the equation of this line $y = -3$?

9. Complete a table of 5 ordered pairs for a horizontal line that passes through the y-axis at 2.

x	y

10. Write the equation of this line.

11. On a coordinate plane graph the line $y = -5$.

12. Write a sentence summarizing what you have learned.

For problems 13-16, use the table to the right.

13. Examine the table of values at the right. What is special about this table?

x	y
2	1
2	4
2	-3
2	0
2	-5

14. On a coordinate plane, graph each ordered pair and draw a line through them. Describe the line that these ordered pairs form.

15. What is the slope of this line?

16. The equation of this line is $x = 2$. Why is the equation of this line $x = 2$?

Cooperative Learning and Algebra 1: Becky Bride
Kagan Publishing • 1 (800) 933-2667 • www.KaganOnline.com

EXPLORING HORIZONTAL AND VERTICAL LINES

For problems 17-20, use the table to the right.

x	y
−4	−3
−4	1
−4	0
−4	2
−4	−5

17. Examine the table of values at the right. What is special about this table?

18. On a coordinate plane, graph each ordered pair and draw a line through them. Describe the line that these ordered pairs form.

19. What is the slope of this line?

20. The equation of this line is $x = -4$. Why is the equation of this line $x = -4$?

21. Write a table of 5 ordered pairs for the vertical line that passes through the x-axis at 6.

x	y

22. Write the equation of this line.

23. On a coordinate plane graph the line $x = -2$.

24. Write a sentence summarizing what you have learned.

ACTIVITY

13 GRAPH MY LINE

Structure: Sage-N-Scribe

Draw the graph for each problem.

1. $2x + y = 4$

2. $4x - 3y = 9$

3. $x - 5y = 10$

4. $y + 3 = 4$

5. $x - 7 = 2$

6. $5x - 2y = 6$

7. $3x + 4y = 12$

8. $9 - x = 7$

9. $5 + y = 3$

10. $x - y = 0$

Answers are on Transparency 4.2.13.

Cooperative Learning and Algebra 1: Becky Bride
Kagan Publishing • 1 (800) 933-2667 • www.KaganOnline.com

13 ANSWERS

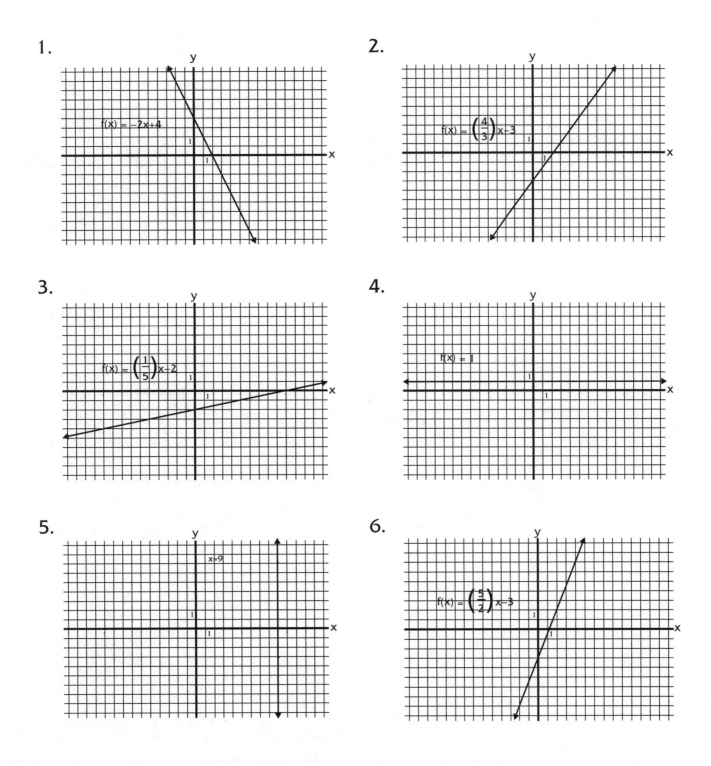

1.

$f(x) = -2x + 4$

2.

$f(x) = \left(\frac{4}{3}\right)x - 3$

3.

$f(x) = \left(\frac{1}{5}\right)x - 2$

4.

$f(x) = 1$

5.

$x = 9$

6.

$f(x) = \left(\frac{5}{2}\right)x - 3$

ACTIVITY

13 ANSWERS

7.

8.

9.

10.

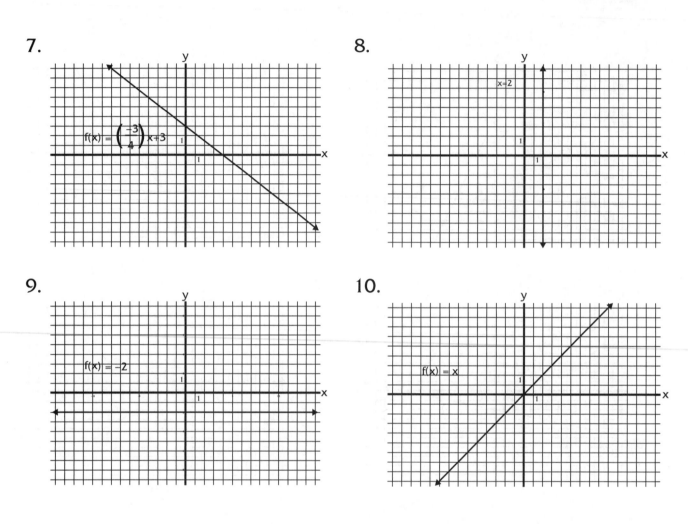

Cooperative Learning and Algebra 1: Becky Bride
Kagan Publishing • 1 (800) 933-2667 • www.KaganOnline.com

LESSON 3
WRITING EQUATIONS OF LINES

Lesson 3 is devoted to writing equations of lines. The activities ask students to write equations of lines in slope-intercept form. Any of these activities could be repeated with point-slope form or standard form. This lesson begins with students writing equations of lines that are graphed, then writing equations when given the slope and y-intercept of a graph, and finally given just two points on the graph. Activity 4 requires higher-level thinking and multiple steps to write the equations of lines given certain criteria. The lesson ends with applications of linear functions and a graphic organizer that sums up the chapter.

ACTIVITY
1 WRITE MY EQUATION, GIVEN A GRAPH

Graphical to Algebraic

Setup:
In pairs, Student A is the Sage; Student B is the Scribe. Students fold a sheet of paper in half and each writes his/her name on one half.

1. The Sage gives the Scribe step-by-step instructions on how to write the equation of the line for problem 1.

2. The Scribe records the Sage's instructions step-by-step in writing.

3. If the Sage is correct, the Scribe praises the Sage. Otherwise, the Scribe coaches, then praises.

4. Students switch roles for the next problem.

Answers:

1. $y = \frac{1}{2}x$

2. $y = \frac{-3}{4}x + 1$

3. $y = 3$

4. $x = 5$

5. $y = \frac{-2}{5}x + 3$

6. $y = \frac{2}{3}x - 4$

▶ **Structure**
• Sage-N-Scribe

▶ **Materials**
• 1 Blackline 4.3.1 per pair
• 1 sheet of paper and pencil per pair

ACTIVITY

2 WRITE MY EQUATION, GIVEN A SLOPE AND A Y-INTERCEPT

▶ **Structure**
• Sage-N-Scribe

▶ **Materials**
• Transparency 4.3.2
• 1 sheet of paper and pencil per pair

Algebraic

Setup:
In pairs, Student A is the Sage; Student B is the Scribe. Students fold a sheet of paper in half and each writes his/her name on one half.

1. The Sage gives the Scribe step-by-step instructions on how to write the equation of the line for problem 1.

2. The Scribe records the Sage's instructions step-by-step in writing.

3. If the Sage is correct, the Scribe praises the Sage. Otherwise, the Scribe coaches, then praises.

4. Students switch roles for the next problem.

ACTIVITY

3 WRITE MY EQUATION, GIVEN TWO POINTS

▶ **Structure**
• Sage-N-Scribe

▶ **Materials**
• Transparency 4.3.3
• 1 sheet of paper and pencil per pair

Algebraic

Setup:
In pairs, Student A is the Sage; Student B is the Scribe. Students fold a sheet of paper in half and each writes his/her name on one half.

1. The Sage gives the Scribe step-by-step instructions on how to write the equation of the line for problem 1.

2. The Scribe records the Sage's instructions step-by-step in writing.

3. If the Sage is correct, the Scribe praises the Sage. Otherwise, the Scribe coaches, then praises.

4. Students switch roles for the next problem.

Cooperative Learning and Algebra 1: Becky Bride
Kagan Publishing • 1 (800) 933-2667 • www.KaganOnline.com

ACTIVITY 4
WRITE THE EQUATION FROM MY RIDDLE

Algebraic

Setup:
In pairs, Student A is the Sage; Student B is the Scribe. Students fold a sheet of paper in half and each writes his/her name on one half.

1. The Sage gives the Scribe step-by-step instructions on how to write the equation of the line for problem 1.

2. The Scribe records the Sage's instructions step-by-step in writing.

3. If the Sage is correct, the Scribe praises the Sage. Otherwise, the Scribe coaches, then praises.

4. Students switch roles for the next problem.

▶ **Structure**
• Sage-N-Scribe

▶ **Materials**
• Transparency 4.3.4
• 1 sheet of paper and pencil per pair

ACTIVITY 5
LINEAR APPLICATIONS

Applications

Setup:
In pairs, Student A is the Sage; Student B is the Scribe. Students fold a sheet of paper in half and each writes his/her name on one half.

1. The Sage gives the Scribe step-by-step instructions on how to solve problem 1.

2. The Scribe records the Sage's instructions step-by-step in writing.

3. If the Sage is correct, the Scribe praises the Sage. Otherwise, the Scribe coaches, then praises.

4. Students switch roles for the next problem.

Answers:

1. a. $C = 15 + 0.39p$
   b. $22.80
   c. 36 people

2. a. $P = 0.50c - 10$
   b. $40
   c. 85 cookies

3. a. $C = 120 + 30m$
   b. 25 months
   c. $480

4. a. $V = 15,000 - 620y$
   b. 6 years
   c. $8,800

5. a. $C = 50 + 0.40m$
   b. 435 miles
   c. $250

6. a. $B = 8000 - 215p$
   b. $4,130
   c. 24 payments

▶ **Structure**
• Sage-N-Scribe

▶ **Materials**
• 1 Blackline 4.3.5 per pair
• 1 sheet of paper and pencil per pair

6 CREATE AN APPLICATION

In this activity, students will create, as a team effort, applications for linear functions. Teammates will define variables and create story problems for their variables through a series of stages. All four teammates begin their own problem so that when the process is finished, four story problems are produced by each team and each teammate has had the opportunity to contribute to each story problem.

▶ **Structure**
· Simultaneous RoundTable

▶ **Materials**
· 1 sheet of paper and pencil per student

Algebraic

1. The teacher tells students to select an object to represent a "unit."

2. Each teammate selects something to represent a "unit" for a story problem and writes it at the top of his/her paper. For example, Teammate 1 may write "minutes." Teammate 2 will likely select a different "unit."

3. The teacher tells students to select another object to represent a "unit."

4. Each teammate selects something to represent another "unit" for a story problem and writes it under the first "unit" on his/her paper. For example, Teammate 2 may write "dogs."

5. After the teacher signals time, or students indicate they are done by placing their thumbs up, they pass the paper one person clockwise.

6. Each teammate reads the "units" their teammate wrote and writes related variables. For the above example, the next student's variable could be "m = total number of minutes and d = total number of dogs."

7. When done, students pass their papers one teammate clockwise.

8. Teammates read the units and variables on their new sheet and write a story problem that can later be written as a linear function. To continue the example, "Andrew has a dog grooming business. On average it takes 75 minutes to groom a dog. If he took a 30-minute lunch, write a linear function to model the total minutes he worked today."

9. When done, students pass their papers one teammate clockwise.

10. Teammates read the story problem and write a linear function to represent the story problem. To continue the dog grooming example, "m = 75d + 30."

11. When done, students pass their papers one teammate clockwise.

12. Teammates check to see if the linear function written on their paper is correct. If the equation is not written correctly, the teammate coaches the student who wrote it incorrectly. Each teammate writes a question that pertains to the story. To continue the example, "If Andrew groomed 8 dogs, how long did he work today (put your answer in hours)?"

13. When done, students pass their papers one teammate clockwise.

14. Teammates answer the question on their paper.

15. When done, students pass their papers one teammate clockwise.

16. Teammates check the solution to the question on their paper for accuracy. If the solution is wrong, then the student coaches the student who solved the equation. Once all the solutions are correct the team praises each other.

Management Tip: If students use different colored pencils and write their names at the top of each paper, it is easier to hold each individual accountable.

WHAT DID WE LEARN?

Synthesis via Graphic Organizer

1. Each teammate signs his/her name in the upper-right corner of the team paper with the color pen/pencil he/she is using.

2. One teammate writes "Linear Functions and Vertical Lines" in the center of the team paper in a rectangle.

3. Teammate 1 shares with the team one core concept he/she learned in the entire unit.

4. The student checks for consensus.

5. The teammates show agreement or lack of agreement with thumbs up or down.

6. If there is agreement, the students celebrate and the teammate records the core concept on the graphic organizer, connecting it with a line to the main idea, "Linear Functions and Vertical Lines." If not, teammates discuss the response until there is agreement and then they celebrate.

7. Play continues with the next student's core concept until all core concepts are exhausted.

8. Repeat steps 3-7, with teammates adding details to each core concept and making bridges between related ideas.

▶ **Structure**
• RoundTable Consensus

▶ **Materials**
• 1 large sheet of paper per team
• different color pen or pencil for each student on the team

ACTIVITY 1 WRITE MY EQUATION, GIVEN A GRAPH

Structure: Sage-N-Scribe

Write the equation of each graph below. The scale of each axis is 1.

1.

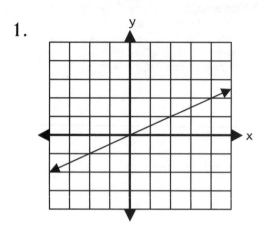

2.

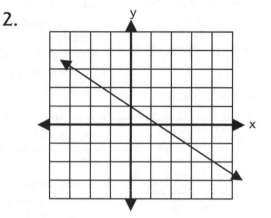

3.

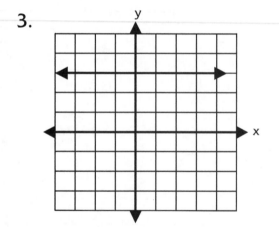

4.

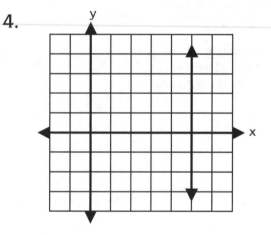

5.

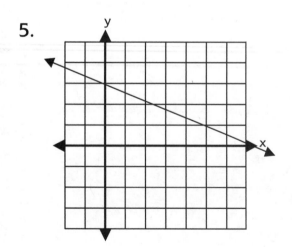

6.

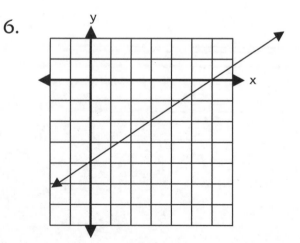

Answers are on the teacher notes (page 273).

Cooperative Learning and Algebra 1: Becky Bride
Kagan Publishing • 1 (800) 933-2667 • www.KaganOnline.com

ACTIVITY 2 — WRITE MY EQUATION, GIVEN A SLOPE AND A Y-INTERCEPT

Structure: Sage-N-Scribe

Write the equation of the line passing through each point with the given slope.

1. $(3, -4)$; $m = \dfrac{2}{3}$

2. $(-1, 6)$; $m = -3$

3. $(5, 7)$; $m = 0$

4. $(-2, -7)$; slope is undefined

5. $(-3, 9)$; $m = \dfrac{-4}{5}$

6. $(8, 3)$; $m = \dfrac{-5}{3}$

Answers:

1. $y = \dfrac{2}{3}x - 6$

2. $y = (-3)x + 3$

3. $y = 7$

4. $x = (-2)$

5. $y = \dfrac{-4}{5}x + \dfrac{33}{5}$

6. $y = \dfrac{-5}{3}x + \dfrac{49}{3}$

WRITE MY EQUATION, GIVEN TWO POINTS

Structure: Sage-N-Scribe

Write the equation of each line passing through the given points.

1. (–5, 1) and (3, 7)

2. (4, –1) and (4, – 8)

3. (7, 5) and (5, –3)

4. (–8, 6) and (–1, –3)

5. (4, 8) and (–15, 8)

6. (12, –3) and (7, –13)

7. (–5, 1) and (9, 1)

8. (–11, 3) and (–11, 5)

Answers:

1. $y = \dfrac{3}{4}x + \dfrac{19}{4}$

2. $x = 4$
3. $y = 4x - 23$

4. $y = \dfrac{-9}{7}x - \dfrac{30}{7}$

5. $y = 8$
6. $y = 2x - 27$
7. $y = 1$
8. $x = (-11)$

Cooperative Learning and Algebra 1: Becky Bride
Kagan Publishing • 1 (800) 933-2667 • www.KaganOnline.com

ACTIVITY

WRITE THE EQUATION
FROM MY RIDDLE

Structure: Sage-N-Scribe

Write the equation of the line, given the following information.

1. The line passes through (–4, 1) and has the same slope as the line whose equation is $4x – 3y = 5$.

2. The line passes through (–3, 7) and has the same y-intercept as the line whose equation is $6x – y = 8$.

3. The line has the same slope as the line $x = –5$ and has the same x-intercept as the line whose equation is $3x – 2y = 6$.

4. The line has the same y-intercept as the line whose equation is $7x + 2y = 14$ and the same x-intercept as the line whose equation is $4y + 3x = 6$.

Answers:

1. $y = \frac{4}{3}x + \frac{19}{3}$

2. $y = (–5)x – 8$

3. $x = 2$

4. $y = \frac{–7}{2}x + 7$

ACTIVITY

5 | LINEAR APPLICATIONS

Structure: Sage-N-Scribe

For each problem:
   a) Write a linear function that models the problem.
   b) Answer the questions using your linear model.

1. Jose is planning a sixteenth birthday party for his friend. The invitations cost $15, and each invitation will cost $0.39 to mail.
   a. Write a linear function if C is the total cost and p is the number of people invited.
   b. What is the total cost if he invites twenty people?
   c. How many people can he invite for $29.04?

2. Jahanna is selling gourmet cookies. She spent $10 on ingredients to make the cookies. She plans to sell the cookies for fifty cents each.
   a. Write a linear function for the profit P, when c cookies are sold.
   b. How much money will Jahanna make if she sells one hundred cookies?
   c. How many cookies did she sell if her profit was $32.50?

3. Shelia wants to join the Get Fit Now gym. The membership fee is $120 and the monthly fee is $30.
   a. Write a linear function if C is the total cost and m is the number of months she goes to the gym.
   b. How many months can she go to the gym for $870?
   c. How much would it cost Shelia if she went to the gym for one year?

Cooperative Learning and Algebra 1: Becky Bride
Kagan Publishing • 1 (800) 933-2667 • www.KaganOnline.com

ACTIVITY

5 **LINEAR APPLICATIONS**

Structure: Sage-N-Scribe

4. Stuart bought a used car at the Nearly New Car store and paid $15,000. The car depreciates $620 each year.
   a. Write a linear function for the value V if he owns the car for y years.
   b. How many years has he owned his car if the value is $11,280?
   c. What is the value of his car after he owns it for ten years?

5. The Moooove-In Truck Rental Company charges a flat fee of fifty dollars, plus an additional forty cents per mile driven, for renting a moving van.
   a. Write a linear function for the total cost C to rent a truck that will be driven m miles.
   b. How many miles can the truck be driven for two hundred twenty-four dollars?
   c. How much will it cost to rent a moving van if it is driven five hundred miles?

6. The Borrow Now Loan company loans Kaylie eight thousand dollars to purchase a car. Kaylie's monthly payment is two hundred fifteen dollars.
   a. Write a linear function for the loan balance B for p monthly payments.
   b. What is the balance of Kaylie's loan after one and one-half years?
   c. How many payments has Kaylie made if her loan balance is two thousand eight hundred forty dollars?

Answers are on teacher notes.

LINEAR SYSTEMS

This chapter works with systems of linear equations and inequalities. Three exploratory activities are included so students can discover what a solution to a linear system of equations or inequalities looks like graphically and the connection between the graphical and algebraic solution of systems with no solution and systems with infinite solutions. Activities are included so students can practice solving systems of linear equations graphically and algebraically—using substitution and elimination methods. Applications are included for both systems of equations and inequalities. Activities to synthesize the lessons are also included.

LESSON 1 LINEAR SYSTEMS

ACTIVITY 1: Exploring Solutions to Linear Systems Graphically
ACTIVITY 2: Find My Solution Graphically
ACTIVITY 3: Am I a Solution?
ACTIVITY 4: Solve My System Using Substitution
ACTIVITY 5: Advanced Substitution
ACTIVITY 6: Beginning Elimination
ACTIVITY 7: Intermediate Elimination
ACTIVITY 8: Advanced Elimination
ACTIVITY 9: Exploring All Possible Solutions with a Graphical-Algebraic Connection
ACTIVITY 10: Elimination—All Possibilities
ACTIVITY 11: Create an Application
ACTIVITY 12: What Did We Learn?

LESSON 2 SYSTEMS OF LINEAR INEQUALITIES

ACTIVITY 1: Exploring Solutions of Linear Inequalities
ACTIVITY 2: Am I a Solution?
ACTIVITY 3: Graph My Linear Inequality
ACTIVITY 4: Graph My System of Linear Inequalities
ACTIVITY 5: Find My Feasible Region
ACTIVITY 6: Linear Programming
ACTIVITY 7: What Did We Learn?

LESSON 1
LINEAR SYSTEMS

This lesson is devoted to systems of linear equations. The students first discover what a solution to a linear system looks like graphically. The first exploration investigates single solution systems only. Activity 9 has students explore the no solution and infinite solution systems. The no solution and infinite solution systems are investigated later in the lesson so the students have the graphical and algebraic means of solving a system of linear equations under their belts. Activities are included that process the substitution method and the elimination method of algebraically finding solutions. These activities progress in difficulty so the students can work with the initial concepts with simple systems. As they feel more comfortable with the mechanics of solving systems, new steps are added. Activity 11 has students, as a team, generate applications of systems of linear equations similar to previous chapters. The lesson ends with a synthesis activity where students create a graphic organizer as a team to summarize what they have learned in the lesson.

ACTIVITY 1
EXPLORING SOLUTIONS TO LINEAR SYSTEMS GRAPHICALLY

The prerequisite skills for this activity include rewriting a linear function in slope-intercept form and graphing linear functions. The goal is to have students see that the numerical representation of a solution of a linear system is an ordered pair, to understand that graphically this solution is the point of intersection, and algebraically the solution satisfies the equations of both linear functions. Only systems that have points of intersection are explored in this investigation. Parallel lines and linear systems with infinite solutions are investigated later in this chapter.

▶ **Structure**
  • Solo-Pair Consensus-Team Consensus

▶ **Materials**
  • 1 Blackline 5.1.1 per student
  • 1 sheet of graph paper and pen/pencil per student

Exploratory

Solo

1. Have each student complete the investigation individually.

Pair Consensus

2. For each problem on the investigation, each student shares with his/her partner, using RallyRobin, his/her response. They discuss the problems that they disagree on, trying to come to consensus on the correct response. For the problems they can't reach consensus on, they mark these so they can focus on them during the team phase. Encourage the students to add to their responses if their partner verbalized an understanding they did not see.

Team Consensus

3. Each pair shares their responses, using RallyRobin, with the other pair in their team, augmenting their responses if necessary. When the teams are through sharing, each student should have a detailed, complete summary.

Cooperative Learning and Algebra 1: Becky Bride
Kagan Publishing • 1 (800) 933-2667 • www.KaganOnline.com

ACTIVITY

ACTIVITY

2

FIND MY SOLUTION GRAPHICALLY

Graphical

Setup:
In pairs, Student A is the Sage; Student B is the Scribe. Students fold a sheet of paper in half and each writes his/her name on one half.

1. The Sage gives the Scribe step-by-step instructions on how to graph problem 1.

2. The Scribe graphs the lines using the Sage's instructions.

3. If the Sage is correct, the Scribe praises the Sage. Otherwise, the Scribe coaches, then praises.

4. Students switch roles for the next problem.

▶ **Structure**
• Sage-N-Scribe

▶ **Materials**
• Transparency 5.1.2
• 1 sheet of graph paper and pencil per pair

ACTIVITY

3

AM I A SOLUTION?

Algebraic

Setup:
In pairs, Student A is the Sage; Student B is the Scribe. Students fold a sheet of paper in half and each writes his/her name on one half.

1. The Sage gives the Scribe step-by-step instructions on how to solve problem 1.

2. The Scribe records the Sage's instructions step-by-step in writing.

3. If the Sage is correct, the Scribe praises the Sage. Otherwise, the Scribe coaches, then praises.

4. Students switch roles for the next problem.

▶ **Structure**
• Sage-N-Scribe

▶ **Materials**
• Transparency 5.1.3
• 1 sheet of paper and pencil per pair

ACTIVITY

4

SOLVE MY SYSTEM USING SUBSTITUTION

Algebraic

Setup:
In pairs, Student A is the Sage; Student B is the Scribe. Students fold a sheet of paper in half and each writes his/her name on one half.

1. The Sage gives the Scribe step-by-step instructions on how to solve problem 1 using the substitution method.

2. The Scribe records the Sage's instructions step-by-step in writing.

3. If the Sage is correct, the Scribe praises the Sage. Otherwise, the Scribe coaches, then praises.

4. Students switch roles for the next problem.

▶ **Structure**
• Sage-N-Scribe

▶ **Materials**
• Transparency 5.1.4
• 1 sheet of paper and pencil per pair

Cooperative Learning and Algebra 1: Becky Bride
Kagan Publishing • 1 (800) 933-2667 • www.KaganOnline.com

ACTIVITY

5 ADVANCED SUBSTITUTION

▶ **Structure**
· Sage-N-Scribe

▶ **Materials**
· Transparency 5.1.5
· 1 sheet of paper and pencil per pair

Algebraic

Setup:
In pairs, Student A is the Sage; Student B is the Scribe. Students fold a sheet of paper in half and each writes his/her name on one half.

1. The Sage gives the Scribe step-by-step instructions on how to solve problem 1 using the substitution method.

2. The Scribe records the Sage's instructions step-by-step in writing.

3. If the Sage is correct, the Scribe praises the Sage. Otherwise, the Scribe coaches, then praises.

4. Students switch roles for the next problem.

ACTIVITY

6 BEGINNING ELIMINATION

▶ **Structure**
· Sage-N-Scribe

▶ **Materials**
· Transparency 5.1.6
· 1 sheet of paper and pencil per pair

Algebraic

Setup:
In pairs, Student A is the Sage; Student B is the Scribe. Students fold a sheet of paper in half and each writes his/her name on one half.

1. The Sage gives the Scribe step-by-step instructions on how to solve problem 1 using the elimination method.

2. The Scribe records the Sage's instructions step-by-step in writing.

3. If the Sage is correct, the Scribe praises the Sage. Otherwise, the Scribe coaches, then praises.

4. Students switch roles for the next problem.

ACTIVITY

7 INTERMEDIATE ELIMINATION

▶ **Structure**
· Sage-N-Scribe

▶ **Materials**
· Transparency 5.1.7
· 1 sheet of paper and pencil per pair

Algebraic

Setup:
In pairs, Student A is the Sage; Student B is the Scribe. Students fold a sheet of paper in half and each writes his/her name on one half.

1. The Sage gives the Scribe step-by-step instructions on how to solve problem 1 using the elimination method.

2. The Scribe records the Sage's instructions step-by-step in writing.

3. If the Sage is correct, the Scribe praises the Sage. Otherwise, the Scribe coaches, then praises.

4. Students switch roles for the next problem.

Cooperative Learning and Algebra 1: Becky Bride
Kagan Publishing • 1 (800) 933-2667 • www.KaganOnline.com

8 ADVANCED ELIMINATION

Algebraic

Setup:
In pairs, Student A is the Sage; Student B is the Scribe. Students fold a sheet of paper in half and each writes his/her name on one half.

1. The Sage gives the Scribe step-by-step instructions on how to solve problem 1 using the elimination method.

2. The Scribe records the Sage's instructions step-by-step in writing.

3. If the Sage is correct, the Scribe praises the Sage. Otherwise, the Scribe coaches, then praises.

4. Students switch roles for the next problem.

▶ **Structure**
• Sage-N-Scribe

▶ **Materials**
• Transparency 5.1.8
• 1 sheet of graph paper and pencil per pair

9 EXPLORING ALL POSSIBLE SOLUTIONS WITH A GRAPHICAL-ALGEBRAIC CONNECTION

Exploratory

Solo

1. Have each student complete the investigation individually.

Pair Consensus

2. For each problem on the investigation, each student shares with his/her partner, using RallyRobin, his/her response. They discuss the problems that they disagree on, trying to come to consensus on the correct response. For the problems they can't

reach consensus on, they mark these so they can focus on them during the team phase. Encourage the students to add to their responses if their partner verbalized an understanding they did not see.

Team Consensus

3. Each pair shares their responses, using RallyRobin, with the other pair in their team, augmenting their responses if necessary. When the teams are through sharing, each student should have a detailed, complete summary.

▶ **Structure**
• Solo-Pair Consensus-Team Consensus

▶ **Materials**
• 1 Blackline 5.1.9 per student
• 1 sheet of graph paper and pen/pencil per student

10 ELIMINATION—ALL POSSIBILITIES

▶ **Structure**
  • Sage-N-Scribe

▶ **Materials**
  • Transparency 5.1.10
  • 1 sheet of paper and pencil per pair

Algebraic

Setup:
In pairs, Student A is the Sage; Student B is the Scribe. Students fold a sheet of paper in half and each writes his/her name on one half.

1. The Sage gives the Scribe step-by-step instructions on how to solve problem 1 using the elimination method.

2. The Scribe records the Sage's instructions step-by-step in writing.

3. If the Sage is correct, the Scribe praises the Sage. Otherwise, the Scribe coaches, then praises.

4. Students switch roles for the next problem.

Cooperative Learning and Algebra 1: Becky Bride
Kagan Publishing • 1 (800) 933-2667 • www.KaganOnline.com

ACTIVITY

11 CREATE AN APPLICATION

This is a fun activity where students will create, as a team effort, applications for linear systems. Teammates will define variables and create story problems for their variables through a series of stages. All four teammates begin their own problem so that when the process is finished, four story problems are produced by each team and each teammate has had the opportunity to contribute to each story problem.

Algebraic

1. The teacher tells students to select an object to represent a "unit."

2. Each teammate selects something to represent a "unit" for a story problem and writes it at the top of his/her paper. For example, Teammate 1 may write "dollars." Teammate 2 will likely select a different "unit."

3. After the teacher signals time, or students indicate they are all done by placing their thumbs up, they pass the paper one person clockwise.

4. The teacher tells students to select another object to represent a "unit."

5. Each teammate selects something to represent another "unit" for a story problem and writes it under the first "unit" on his/her paper. For example, teammate 2 may write "months."

6. After the teacher signals time, or students indicate they are all done by placing their thumbs up, they pass the paper one person clockwise.

7. Each teammate reads the "units" their teammates wrote and writes related variables. For the above example, the

next student's variable could be "C = total cost in dollars and m = number of months."

8. When done, students pass their papers one teammate clockwise.

9. Teammates read the units and variables on their new sheet and write a story problem that can later be written with 2 linear functions. To continue the example, "Paintball Jungle has two different types of memberships. One type of membership requires $50 to join and $20 a month to keep the membership active while the second type requires $10 to join and $25 a month to keep the membership active."

10. When done, students pass their papers one teammate clockwise.

11. Teammates read the story problem and write a linear function to represent the story problem. To continue the paintball example, "$C = 20d + 50$ and $C = 25d + 10$."

12. When done, students pass their papers one teammate clockwise.

13. Teammates check to see if the two linear functions written on their paper are correct. If the equations are not written correctly, the teammate coaches the student who wrote it

> ▶ **Structure**
> · Simultaneous RoundTable
>
> ▶ **Materials**
> · 1 sheet of paper and pencil per student

incorrectly. Each teammate writes a question that pertains to the story. To continue the example, "Juan bought the first type of membership and Elisa bought the second type of membership. How many months will it take for Juan and Elisa to have paid the same amount of money?"

14. When done, students pass their papers one teammate clockwise.

15. Teammates answer the question on their paper.

16. When done, students pass their papers one teammate clockwise.

17. Teammates check the solution to the question on their paper for accuracy. If the solution is wrong, then the student coaches the student who solved the equation. Once all the solutions are correct the team praises each other.

Management Tip: If students use different colored pencils and write their names at the top of each paper, it is easier to hold each individual accountable.

ACTIVITY

12 WHAT DID WE LEARN?

▶ **Structure**
- RoundTable Consensus

▶ **Materials**
- 1 large sheet of paper per team
- different color pen or pencil for each student on the team

Synthesis via Graphic Organizer

1. Each teammate signs his/her name in the upper right corner of the team paper with the color pen/pencil he/she is using.

2. One teammate writes "Linear Systems" in the center of the team paper in a rectangle.

3. Teammate 1 shares with the team one core concept he/she learned in the unit.

4. The student checks for consensus.

5. The teammates show agreement or lack of agreement with thumbs up or down.

6. If there is agreement, the students celebrate and the teammate records the core concept on the graphic organizer connecting it with a line to the main idea, "Linear Systems." If not, teammates discuss the response until there is agreement and then they celebrate.

7. Play continues with the next student's core concept until all core concepts are exhausted.

8. Repeat steps 3-7, with teammates adding details to each core concept and making bridges between related ideas.

Cooperative Learning and Algebra 1: Becky Bride
Kagan Publishing • 1 (800) 933-2667 • www.KaganOnline.com

ACTIVITY 1 EXPLORING SOLUTIONS TO LINEAR SYSTEMS GRAPHICALLY

For the first part of the investigation, use the following linear functions:

$x + 2y = 10$

$-x + 4y = 8$

The two equations above form what mathematicians call a "linear system of equations." The purpose of this investigation is to explore the solution to a linear system.

1. Graph the first function.

2. On the same coordinate plane, graph the second function.

3. Write the coordinates of the point of intersection to the right.

4. This point is called the **"solution"** to the linear system. Describe how you would determine the solution of a linear system if you were given a graph of that system.

5. What is the solution to the system graphed below?

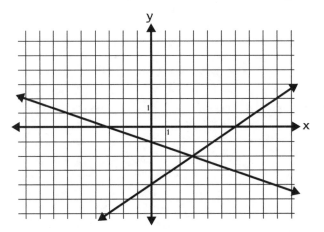

EXPLORING SOLUTIONS TO LINEAR SYSTEMS GRAPHICALLY

6. To understand how to tell algebraically if an ordered pair is a solution to a linear system, substitute the ordered pair (4, 3) (which was the solution of the system you investigated in problems 1-4) into the appropriate variables for each equation below. Show your work.

$x + 2y = 10$ $\qquad\qquad$ $-x + 4y = 8$

7. What was special about the result from each equation?

8. To clarify this further,

a. Looking at the graph, is the point (2, 4) on line 1?

b. Looking at the graph, is the point (2, 4) on line 2?

c. Since (2, 4) is not on both lines, it is not a solution.

d. Substitute the coordinates of (2, 4) into the appropriate variables for each equation below like you did in problem 6. Show your work.

$x + 2y = 10$ $\qquad\qquad$ $-x + 4y = 8$

9. What is different about the results from this ordered pair compared to the results in problem 6?

Cooperative Learning and Algebra 1: Becky Bride
Kagan Publishing • 1 (800) 933-2667 • www.KaganOnline.com

EXPLORING SOLUTIONS TO LINEAR SYSTEMS GRAPHICALLY

10. Based on your answers to problems 6-9, describe how to tell algebraically if an ordered pair is a solution to a linear system.

11. Use the new knowledge described in problem 10 to determine which ordered pair below is a solution to the following system. Show your work.

    $2x - 3y = -12$ $4x + 3y = -6$

    a. (3, 6)   b. (−3, 2)

12. Graph each line in problem 11 to verify your answers.

13. Summarize what it means to be a solution to a linear system of equations graphically and algebraically.

ACTIVITY
2 FIND MY SOLUTION GRAPHICALLY

Structure: Sage-N-Scribe

Find the solution of each system by graphing.

1. $-x + 3y = 6$
   $2x + 3y = -3$

2. $y - 3x = 0$
   $2x + y = 5$

3. $-x + 2y = -6$
   $x + 2y = 2$

4. $5y + x = 5$
   $5y - 3x = 25$

Answers:
1. $(-3, 1)$
2. $(1, 3)$
3. $(4, -1)$
4. $(-5, 2)$

Cooperative Learning and Algebra 1: Becky Bride
Kagan Publishing • 1 (800) 933-2667 • www.KaganOnline.com

3 AM I A SOLUTION?

Structure: Sage-N-Scribe

Determine if the ordered pair is a solution of the given system.

1. (–4, 2) $4x - 3y = (-22)$
               $2x + y = (-6)$

2. (3, –7) $x - 5y = 38$
               $3x + 4y = (-20)$

3. (–2, 3) $7y = 2x + 25$
               $3x = 16 - 2y$

4. (5, 1) $3y = 4x - 17$
               $2x + y = 11$

Answers:
1. yes
2. no
3. no
4. yes

ACTIVITY 4

SOLVE MY SYSTEM USING SUBSTITUTION

Structure: Sage-N-Scribe

Solve each system using the substitution method.

1. $2x - 3y = -11$
   $x = 3y - 13$

2. $4x + 7y = 15$
   $y = 6x - 11$

3. $5x + 3y = 3$
   $y = 2x + 12$

4. $4x - 3y = 7$
   $x = 8 + 2y$

Answers:
1. (2, 5)
2. (2, 1)
3. (−3, 6)
4. (−2, −5)

Cooperative Learning and Algebra 1: Becky Bride
Kagan Publishing • 1 (800) 933-2667 • www.KaganOnline.com

ADVANCED SUBSTITUTION

Structure: Sage-N-Scribe

Solve each system using the substitution method.

1. $2x - y = (-20)$
   $3x + 4y = (-8)$

2. $x + 5y = 18$
   $2x - 3y = (-3)$

3. $4x + 5y = 31$
   $x + 7y = 48$

4. $2x + y = (-1)$
   $3x - y = 11$

Answers:
1. (–8, 4)
2. (3, 3)
3. (–1, 7)
4. (2, –5)

BEGINNING ELIMINATION

Structure: Sage-N-Scribe

Solve each system using the elimination method.

1. $2x - 5y = (-31)$
   $3x + 5y = 16$

2. $4x - 7y = 31$
   $4x + 3y = 21$

3. $6x - 5y = (-37)$
   $6x + 2y = (-2)$

4. $3x - y = (-7)$
   $2x + y = 7$

Answers:
1. (−3, 5)
2. (6, −1)
3. (−2, 5)
4. (0, 7)

Cooperative Learning and Algebra 1: Becky Bride
Kagan Publishing • 1 (800) 933-2667 • www.KaganOnline.com

INTERMEDIATE ELIMINATION

Structure: Sage-N-Scribe

Solve each system using the elimination method.

1. $2x + 5y = (-24)$
   $3x + y = (-10)$

2. $4x - 7y = 3$
   $2x + 8y = 36$

3. $9x - 8y = 64$
   $5x + 2y = 42$

4. $9y - 10x = 104$
   $4y - 2x = 34$

Answers:
1. $(-2, -4)$
2. $(6, 3)$
3. $(8, 1)$
4. $(-5, 6)$

ADVANCED ELIMINATION

Structure: Sage-N-Scribe

Solve each system using the elimination method.

1. $3x - 5y = (-7)$
   $2x + 3y = 8$

2. $4x + 3y = 5$
   $5x + 2y = 1$

3. $7x - 2y = 20$
   $3x - 5y = 21$

4. $8x - 3y = (-13)$
   $5x + 7y = (-17)$

Answers:
1. (1, 2)
2. (−1, 3)
3. (2, −3)
4. (−2, −1)

Cooperative Learning and Algebra 1: Becky Bride
Kagan Publishing • 1 (800) 933-2667 • www.KaganOnline.com

ACTIVITY 9 EXPLORING ALL POSSIBLE SOLUTIONS WITH A GRAPHICAL-ALGEBRAIC CONNECTION

1. Graph the linear system of equations below on the same coordinate plane.

    $-x + 2y = 4$

    $-2x + 4y = 12$

2. From the last exploration, you learned that the solution to a linear system was the point(s) of intersection. What is the solution of this system? **EXPLAIN.**

3. Use the elimination method to solve the linear system in problem 1. Show your work below.

4. When you added the two equations, what happened to the variables x and y?

5. Is the final equation after you finished solving this system a true or false statement?

EXPLORING ALL POSSIBLE SOLUTIONS WITH A GRAPHICAL-ALGEBRAIC CONNECTION

6. Graph the linear system of equations below on the same coordinate plane.
   $6x - 2y = 8$
   $9x - 3y = 12$

7. What is the solution of this system? **EXPLAIN.**

8. Use the elimination method to solve the linear system in problem 6. Show your work below.

9. When you added the two equations, what happened to the variables x and y?

10. Is the final equation after you finished solving this system a true or false statement?

11. Compare and contrast the algebraic results from problems 3 and 8. Explain how you will know which system has infinite solutions and which has no solution when you solve a linear system algebraically.

Cooperative Learning and Algebra 1: Becky Bride
Kagan Publishing • 1 (800) 933-2667 • www.KaganOnline.com

Structure: Sage-N-Scribe

Solve each system using the elimination method.

1. $5x + 3y = 4$
   $-10x - 6y = -8$

2. $8x + 4y = (-12)$
   $10x + 6y = (-8)$

3. $9x + 6y = 11$
   $-12x - 8y = 5$

4. $7x - 3y = 6$
   $-21x + 9y = 18$

Answers:

1. $\left\{ x \mid x = \dfrac{4}{5} - \dfrac{3}{5}y \right\}$

2. $(-5, 7)$

3. no solution

4. no solution

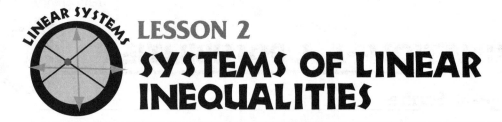

LESSON 2
SYSTEMS OF LINEAR INEQUALITIES

This lesson is devoted to systems of linear inequalities. The first exploratory activity has students discover the meaning of a solution to a system of inequalities. This discovery will help students understand why the test point is needed to determine which side of the line to shade for a linear inequality, and students will see why the answers to these systems are shaded regions. Students will practice how to graph a single linear inequality on the coordinate plane before they learn how to solve systems of linear inequalities. The next activity requires the students to graph a linear system of inequalities to find its solution. The activity for the feasible region develops prerequisite skills for the linear programming activity that follows. The lesson wraps up with a synthesis activity, bringing all the concepts together.

1 EXPLORING SOLUTIONS OF LINEAR INEQUALITIES

▶ **Structure**
· RallyTable

▶ **Materials**
· 1 Blacklines 5.2.1a and 5.2.1b per pair
· 1 pencil per student
· 1 highlighter per pair

Exploratory

1. Teacher gives each pair Blackline 5.2.1a and 5.2.1b.

2. In pairs, students take turns passing the paper and pencil, each writing one answer or making a contribution.

2 AM I A SOLUTION?

▶ **Structure**
· Sage-N-Scribe

▶ **Materials**
· Transparency 5.2.2
· 1 sheet of paper and pencil per pair

Graphical

Setup:
In pairs, Student A is the Sage; Student B is the Scribe. Students fold a sheet of paper in half and each writes his/her name on one half.

1. The Sage gives the Scribe step-by-step instructions on how to determine if the ordered in problem 1 is a solution to the inequality.

2. The Scribe records the Sage's instructions step-by-step in writing.

3. If the Sage is correct, the Scribe praises the Sage. Otherwise, the Scribe coaches, then praises.

4. Students switch roles for the next problem.

ACTIVITY 3
GRAPH MY LINEAR INEQUALITY

Graphical

Setup:
In pairs, Student A is the Sage; Student B is the Scribe. Students fold a sheet of paper in half and each writes his/her name on one half.

1. The Sage gives the Scribe step-by-step instructions on how to graph problem 1.

2. The Scribe graphs the inequality using the Sage's instructions.

3. If the Sage is correct, the Scribe praises the Sage. Otherwise, the Scribe coaches, then praises.

4. Students switch roles for the next problem.

▶ **Structure**
• Sage-N-Scribe

▶ **Materials**
• Transparency 5.2.3
• Answer Transparency 5.2.3a
• 1 sheet of graph paper and pencil per pair

ACTIVITY 4
GRAPH MY SYSTEM OF LINEAR INEQUALITIES

Graphical

Setup:
In pairs, Student A is the Sage; Student B is the Scribe. Students fold a sheet of paper in half and each writes his/her name on one half.

1. The Sage gives the Scribe step-by-step instructions on how to graph problem 1.

2. The Scribe graphs the inequality using the Sage's instructions.

3. If the Sage is correct, the Scribe praises the Sage. Otherwise, the Scribe coaches, then praises.

4. Students switch roles for the next problem.

▶ **Structure**
• Sage-N-Scribe

▶ **Materials**
• Transparency 5.2.4
• Answer Transparency 5.2.4a
• 1 sheet of graph paper and pencil per pair

ACTIVITY 5
FIND MY FEASIBLE REGION

Graphical

Setup:
In pairs, Student A is the Sage; Student B is the Scribe. Students fold a sheet of paper in half and each writes his/her name on one half.

1. The Sage gives the Scribe step-by-step instructions on how to graph problem 1 and what to shade for the feasible region.

2. The Scribe graphs the feasible region using the Sage's instructions.

3. If the Sage is correct, the Scribe praises the Sage. Otherwise, the Scribe coaches, then praises.

4. Students switch roles for the next problem.

▶ **Structure**
• Sage-N-Scribe

▶ **Materials**
• Transparency 5.2.5
• Answer Transparency 5.2.5a
• 1 sheet of graph paper and pencil per pair

Chapter 5: Linear Systems
Lesson Two

LINEAR PROGRAMMING

▶ **Structure**
• Sage-N-Scribe

▶ **Materials**
• Blackline 5.2.6
• 1 sheet of paper and pencil per pair

Graphical and Algebraic

Setup:
In pairs, Student A is the Sage; Student B is the Scribe. Students fold a sheet of paper in half and each writes his/her name on one half.

1. The Sage gives the Scribe step-by-step instructions on how to solve problem 1.

2. The Scribe records the Sage's instructions step-by-step in writing.

3. If the Sage is correct, the Scribe praises the Sage. Otherwise, the Scribe coaches, then praises.

4. Students switch roles for the next problem.

WHAT DID WE LEARN?

▶ **Structure**
• RoundTable Consensus

▶ **Materials**
• 1 large sheet of paper per team
• different color pen or pencil for each student on the team

Synthesis via Graphic Organizer

1. Each teammate signs his/her name in the upper right corner of the team paper with the color pen/pencil he/she is using.

2. One teammate writes "Systems of Linear Inequalities" in the center of the team paper in a rectangle.

3. Teammate 1 shares with the team one core concept he/she learned in the unit.

4. The student checks for consensus.

5. The teammates show agreement or lack of agreement with thumbs up or down.

6. If there is agreement, the students celebrate and the teammate records the core concept on the graphic organizer, connecting it with a line to the main idea, "Systems of Linear Inequalities." If not, teammates discuss the response until there is agreement and then they celebrate.

7. Play continues with the next student's core concept until all core concepts are exhausted.

8. Repeat steps 3-7, with teammates adding details to each core concept and making bridges between related ideas.

Cooperative Learning and Algebra 1: Becky Bride
Kagan Publishing • 1 (800) 933-2667 • www.KaganOnline.com

ACTIVITY 1 EXPLORING SOLUTIONS OF LINEAR INEQUALITIES

1. On the coordinate plane, Student A will label the x-axis from –4 to 4 and Student B will will label the y-axis from –4 to 4.

2. Simultaneously, Student A will write the coordinates of each ordered pair in quadrants 2 and 3 and Student B will write the coordinates of each ordered pair in quadrants 1 and 4 very small, just to the left of the point on the coordinate plane. See diagram.

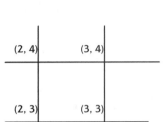

3. Simultaneously, Student A will write the value of x + y for each set of ordered pairs in quadrants 2 and 3 and Student B will write the value of x + y for each set of ordered pairs in quadrants 1 and 4, writing the sum to the right of each ordered pair. See diagram below.

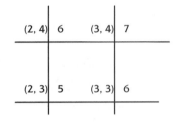

4. Taking turns, the students will circle each point that satisfies the inequality x + y ≤ 4.

5. Discuss with your partner what you notice about all the points that are circled. Student A will write a sentence or two summarizing the discussion.

6. Find the point (3.5, –1.5). Discuss with your partner if this point is a solution of the inequality x + y ≤ 4. Student B will write a sentence or two below summarizing the discussion.

ACTIVITY 1 EXPLORING SOLUTIONS OF LINEAR INEQUALITIES

7. Find the point (5, 1.5). Discuss if this point is a solution of the inequality x + y ≤ 4. Student A will write a sentence or two below summarizing the discussion.

8. Student B will shade on the graph, with a highlighter, the part of the coordinate plane that has all the ordered pairs that are solutions to the inequality x + y ≤ 4.

9. Discuss the shape of the boundary between the shaded area and the unshaded area. Student A will write below a few sentences summarizing the discussion.

10. Discuss how you could algebraically determine (without using the graph) whether the point whose coordinates are (−2, −6) is a solution of the inequality x + y ≤ 4. Student B will write a few sentences below summarizing the discussion or will demonstrate how to algebraically determine if (−2, −6) is a solution.

Cooperative Learning and Algebra 1: Becky Bride
Kagan Publishing • 1 (800) 933-2667 • www.KaganOnline.com

EXPLORING SOLUTIONS OF
LINEAR INEQUALITIES

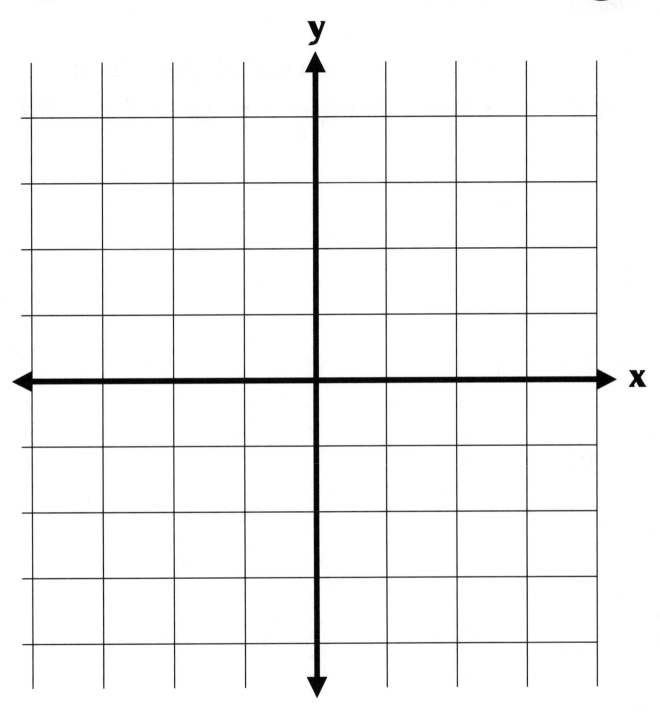

AM I A SOLUTION?

Structure: Sage-N-Scribe

Determine if each ordered pair is a solution of the inequality 2x – 3y > 10.

1. (–3, 5)

2. (8, –2)

3. (4, –1)

4. (8, 2)

Answers:
1. no
2. yes
3. yes
4. no

Cooperative Learning and Algebra 1: Becky Bride
Kagan Publishing • 1 (800) 933-2667 • www.KaganOnline.com

GRAPH MY LINEAR INEQUALITY

Structure: Sage-N-Scribe

Graph each linear inequality on a coordinate plane.

1. $y > 3x - 2$

2. $y \leq 3 - \dfrac{1}{2}x$

3. $x \geq 5$

4. $y + 2 < -1$

ACTIVITY

3

GRAPH MY LINEAR INEQUALITY

Answers

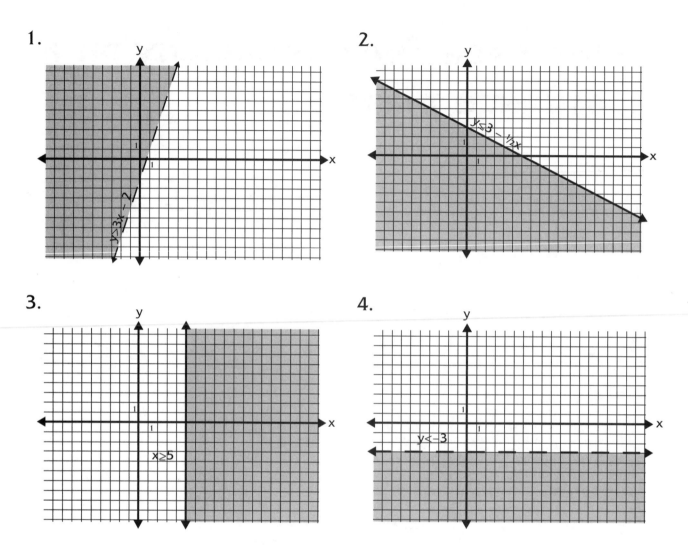

1.

$y \geq 3x - 2$

2.

$y \leq 3 - \frac{1}{2}x$

3.

$x \geq 5$

4.

$y < -3$

Cooperative Learning and Algebra 1: Becky Bride
Kagan Publishing • 1 (800) 933-2667 • www.KaganOnline.com

GRAPH MY SYSTEM OF LINEAR INEQUALITIES

Structure: Sage-N-Scribe

Graph each system of linear inequalities.

1. $x \geq (-2)$
   $y < 2x + 1$

2. $y \geq 3x - 2$
   $y \geq (-3)$

3. $y - 2x < (-3)$
   $2x + 3y > 3$

4. $4y - x > 8$
   $5x - 2y > 2$

GRAPH MY SYSTEM OF LINEAR INEQUALITIES

Answers

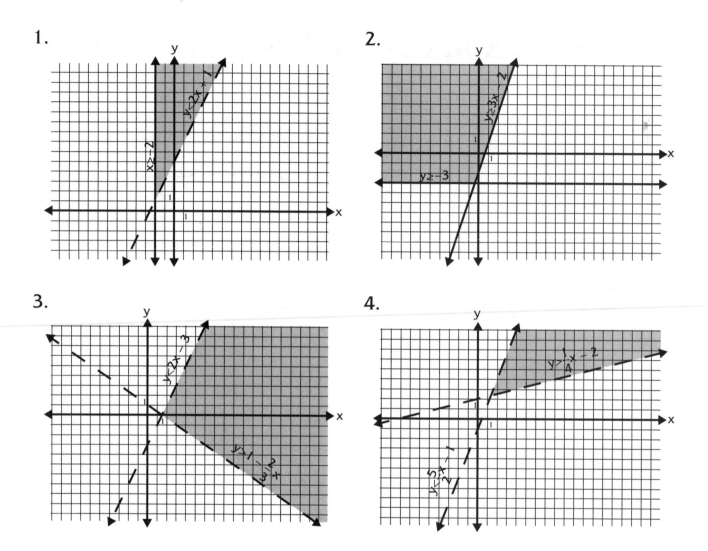

1.

2.

3.

4.

Cooperative Learning and Algebra 1: Becky Bride
Kagan Publishing • 1 (800) 933-2667 • www.KaganOnline.com

ACTIVITY 5 FIND MY FEASIBLE REGION

Structure: Sage-N-Scribe

Graph each system of inequalities and find the feasible region.

1. $y \leq 8 - \dfrac{3}{4}x$

   $x \geq 1$

   $y \geq 2$

2. $x \geq 0$

   $y \geq 0$

   $y \leq 9 - \dfrac{1}{2}x$

   $y \geq 4 - 2x$

3. $x \geq 0$

   $y \geq -3$

   $y \leq 8 - x$

   $y \leq x - 2$

4. $y \geq 3 - x$

   $y \leq 2x + 3$

   $y \leq 10 - x$

   $y \geq 2x - 6$

ACTIVITY
5 FIND MY FEASIBLE REGION

LINEAR SYSTEMS

Answers

1.

2.

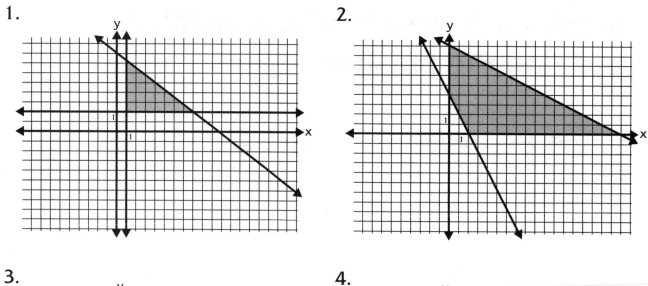

3.

4.

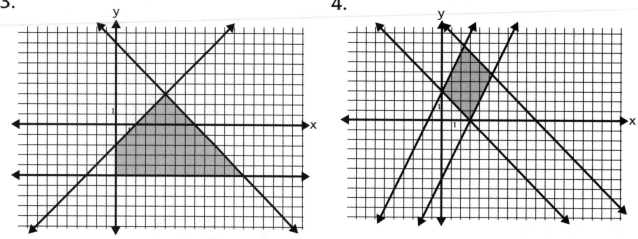

Cooperative Learning and Algebra 1: Becky Bride
Kagan Publishing • 1 (800) 933-2667 • www.KaganOnline.com

LINEAR PROGRAMMING

Structure: Sage-N-Scribe

Solve each problem.

1. The Cookie Emporium makes and sells chocolate chip and gingerbread cookies. They make at most a combined total of 500 cookies each day. They produce at least 100 chocolate chip cookies and at least 100 gingerbread cookies. They make $0.50 on each chocolate chip cookie they sell and $0.40 on each gingerbread cookie they sell. Find the maximum profit and the number of each type of cookie that must be made to yield the maximum profit.

2. The Surf's Up Surfboard Company sells two types of surfboards—the rugged model and the sleek model. The company stocks at least 10 but no more than 20 rugged surfboards. They stock at least 4 sleek surfboards. The number of rugged surfboards is greater than or equal to 15 more than the number of sleek surfboards. Ten times the number of sleek surfboards is less than or equal to 50 more than 7 times the number of rugged surfboards. If the company makes $35 on each rugged surfboard and $30 on each sleek surfboard, find the number of each type of surfboard that must be sold to maximize their profit.

3. The Natural Necklace Company makes and sells necklaces. For one type of necklace, the company uses animal beads and oval beads. The number of animal beads per necklace must be less than or equal to eight beads and there must be at least four oval beads per necklace. Three times the number of oval beads is less than or equal to eight more than twice the number of animal beads. Each animal bead costs $0.20, and each oval bead costs $0.10. Find the minimum cost to make a necklace with animal and oval beads, and find the number of animal and oval beads that yield the minimum cost.

ACTIVITY

6 LINEAR PROGRAMMING

Structure: Sage-N-Scribe

4. The freshman class is selling T-shirts and sweatshirts as a fundraiser. They want to sell at least 20 but not more than 100 T-shirts and at least 20 but not more than 80 sweatshirts. The number of T-shirts they can afford to buy is less than or equal to 60 less than twice the number of sweatshirts. If each T-shirt costs them $6 and each sweatshirt costs them $14, find the number of T-shirts and sweatshirts they should buy to minimize the cost.

Answers:
1. 400 chocolate chip cookies and 100 sugar cookies for a profit of $240
2. 20 rugged surfboards and 19 sleek surfboards for a profit of $1,450
3. 2 animal beads and 4 oval beads for a cost of $0.80
4. 20 T-shirts and 20 sweatshirts for a cost of $390

Cooperative Learning and Algebra 1: Becky Bride
Kagan Publishing • 1 (800) 933-2667 • www.KaganOnline.com

POLYNOMIALS

This chapter begins with students generating definitions specific to polynomials. The remaining activities in Lesson 1 process those definitions. Activity 8 sets in place prerequisite skills for subtracting polynomials in Lesson 2. Lesson 2 is filled with activities to process the operations on polynomials. Applications of these operations are included in Activity 6. Lesson 3 explores and processes factoring. Solving quadratic equations using factoring is in this lesson and will be also covered in Chapter 8. Applications of quadratic equations are included. The chapter ends with an activity to synthesize what students have learned.

LESSON 1 DEVELOPING VOCABULARY

ACTIVITY 1: Define Me
ACTIVITY 2: Am I a Polynomial?
ACTIVITY 3: Name the Degree of Each Term
ACTIVITY 4: Name the Degree of the Polynomial
ACTIVITY 5: Name the Coefficient of Each Term
ACTIVITY 6: Name the Leading Coefficient
ACTIVITY 7: Rewrite Me in Standard Form
ACTIVITY 8: Find the Opposite
ACTIVITY 9: Name My Type
ACTIVITY 10: What Have We Learned?

LESSON 2 OPERATIONS ON POLYNOMIALS

ACTIVITY 1: Add Me
ACTIVITY 2: Subtract Me
ACTIVITY 3: Multiply Me
ACTIVITY 4: Multiply My Binomials
ACTIVITY 5: Multiply My Polynomials
ACTIVITY 6: Apply Me
ACTIVITY 7: Simplify My Division
ACTIVITY 8: Exploring Completing the Square
ACTIVITY 9: Complete My Square

LESSON 3 FACTORING AND SOLVING POLYNOMIAL EQUATIONS

ACTIVITY 1: What Is There in Common?
ACTIVITY 2: Common Factor
ACTIVITY 3: Exploring Trinomial Factoring
ACTIVITY 4: Factor My Trinomial
ACTIVITY 5: Exploring the Difference of Two Squares
ACTIVITY 6: Factoring the Difference of Two Squares
ACTIVITY 7: Factor Me Completely
ACTIVITY 8: Find My Solution
ACTIVITY 9: Solve Me
ACTIVITY 10: Polynomial Applications
ACTIVITY 11: What Have We Learned?

LESSON 1
DEVELOPING VOCABULARY

Lesson 1 is devoted to vocabulary development and processing. Processing each word as it is defined will reinforce the definition and add variety to the lesson. The standard form activity involves one variable only. Many of the activities in Lesson 2 require answers in standard form to reinforce the concept and make grading the polynomials easier. The activity "Find the Opposite" lays the foundation for subtraction of polynomials.

1 DEFINE ME

▶ **Structure**
· Solo Pair Team

▶ **Materials**
· 1 Blackline 6.1.1 per student
· 1 sheet of paper and pencil per student (Solo)
· 1 sheet of paper per team (Team)

Exploratory

Solo

1. Individually, each student defines each vocabulary word by comparing and contrasting the examples and counterexamples. (Students could do this part for homework.)

Team with RoundRobin sharing and RoundTable recording.

2. In turn, each student reads to his/her team the definition he/she wrote.

3. The team discusses how to define the vocabulary word, based on the definitions just shared. The team must come to consensus on how to define the term.

4. Once the team reaches consensus, student 1 records the team definition on the team paper.

5. Repeat steps 2-4 for each vocabulary word, rotating the recording.

Class

6. Choose a team and a student at that team to read his/her team's definition of the first vocabulary word. (Team and student spinners work well here.)

7. Write the definition on the board. Ask the class if they agree with part 1, then part 2, and finally part 3 of the definition, reworking what they want to change.

8. Once everyone has agreed, including the teacher, then it is recorded as the definition the class will use. You may find that lower-level classes prefer a somewhat longer definition if it has more meaning for them. Honors classes will try to be as concise as possible.

9. Repeat steps 6-8 for the remainder of the words.

Chapter 6: Polynomials
Lesson One

322

POLYNOMIALS

ACTIVITY

AM I A POLYNOMIAL?

Algebraic

1. Teacher poses multiple problems with Transparency 6.1.2.

2. In pairs, students take turns stating whether the expression is or is not a polynomial, orally.

▶ **Structure**
• RallyRobin

▶ **Materials**
• Transparency 6.1.2

ACTIVITY

NAME THE DEGREE OF EACH TERM

Algebraic

1. Teacher poses multiple problems with Transparency 6.1.2.

2. In pairs, students take turns stating the degree of each term, orally.

▶ **Structure**
• RallyRobin

▶ **Materials**
• Transparency 6.1.2

ACTIVITY

NAME THE DEGREE OF THE POLYNOMIAL

Algebraic

1. Teacher poses multiple problems with Transparency 6.1.2.

2. In pairs, students take turns stating the degree of the polynomial, orally.

▶ **Structure**
• RallyRobin

▶ **Materials**
• Transparency 6.1.2

ACTIVITY

NAME THE COEFFICIENT OF EACH TERM

Algebraic

1. Teacher poses multiple problems with Transparency 6.1.2.

2. In pairs, students take turns stating the coefficient of each term, orally.

▶ **Structure**
• RallyRobin

▶ **Materials**
• Transparency 6.1.2

Chapter 6: Polynomials Lesson One

ACTIVITY

6

NAME THE LEADING COEFFICIENT

▶ **Structure**
• RallyRobin

▶ **Materials**
• Transparency 6.1.2

Algebraic

1. Teacher poses multiple problems with Transparency 6.1.2.

2. In pairs, students take turns stating the leading coefficient of the polynomial, orally.

ACTIVITY

7

REWRITE ME IN STANDARD FORM

▶ **Structure**
• RallyCoach

▶ **Materials**
• Transparency 6.1.2
• 1 sheet of paper and 1 pencil per pair

Algebraic

1. Teacher poses multiple problems with Transparency 6.1.2.

2. Partner A writes the polynomial in problem 1 in standard form.

3. Partner B watches and listens, checks, and praises.

4. Partner B writes the next polynomial in standard form.

5. Partner A watches and listens, checks, and praises.

6. Repeat for remaining problems starting at step 2.

ACTIVITY

8

FIND THE OPPOSITE

▶ **Structure**
• RallyCoach

▶ **Materials**
• Transparency 6.1.2
• 1 sheet of paper and 1 pencil per pair

Algebraic

1. Teacher poses multiple problems with Transparency 6.1.2.

2. Partner A writes the opposite of the first polynomial.

3. Partner B watches and listens, checks, and praises.

4. Partner B writes the opposite of the next polynomial.

5. Partner A watches and listens, checks, and praises.

6. Repeat for remaining problems starting at step 2.

Cooperative Learning and Algebra 1: Becky Bride
Kagan Publishing • 1 (800) 933-2667 • www.KaganOnline.com

ACTIVITY
9

NAME MY TYPE

Algebraic

1. Teacher poses multiple problems with Transparency 6.1.3.

2. In pairs, students take turns stating the type of the polynomial, orally.

▶ **Structure**
· RallyRobin

▶ **Materials**
· Transparency 6.1.3

ACTIVITY
10

WHAT HAVE WE LEARNED?

Synthesis via Graphic Organizer

1. Each teammate signs his/her name in the upper-right corner of the team paper with the color pen/pencil he/she is using.

2. One teammate writes "Polynomials" in the center of the team paper in a rectangle.

3. Teammate 1 shares with the team one core concept he/she learned in the unit.

4. The student checks for consensus.

5. The teammates show agreement or lack of agreement with thumbs up or down.

6. If there is agreement, the students celebrate and the teammate records the core concept on the graphic organizer, connecting it with a line to the main idea, "Polynomials." If not, teammates discuss the response until there is agreement and then they celebrate.

7. Play continues with the next student's core concept until all core concepts are exhausted.

8. Repeat steps 3-7, with teammates adding details to each core concept and making bridges between related ideas.

▶ **Structure**
· RoundTable Consensus

▶ **Materials**
· 1 large sheet of paper per team
· different color pen or pencil for each student on the team

ACTIVITY

1 DEFINE ME

Structure: Solo Pair Team
For each word, compare the examples with the counterexamples and write a definition.

1. Polynomial

Examples	Counterexamples
$2x + 3x^2 - 4$	$x^{-3} + 3x - 1$
$5xy$	$x^{\frac{2}{3}}$
$-8x^7 - 2x^5 + 3x - 1$	$\dfrac{5}{x} + x^3 - 5$
$-x^3 + 1.5x^4 - \sqrt{6}$	$\sqrt{x} - x^2 + 3x^{2.6}$
$\dfrac{5x^3}{4} + 3x^8 - 7$	$\dfrac{5}{4x^3} + 3x^{-8} - 7$

2. Degree of a term

Examples	Counterexamples

$$4x^2 + 8x - 7$$

Examples	Counterexamples
2 is the degree of the first term	4 is not the degree of the first term
1 is the degree of the second term	8 is not the degree of the second term
0 is the degree of the third term	−7 is not the degree of the third term

$$x^6 + 2x^5 + 3x - 4$$

Examples	Counterexamples
6 is the degree of the first term	1 is not the degree of the first term
5 is the degree of the second term	2 is not the degree of the second term
1 is the degree of the third term	3 is not the degree of the third term
0 is the degree of the fourth term	−4 is not the degree of the fourth term

$$5 + 3x - 6x^2 - x^4$$

Examples	Counterexamples
0 is the degree of the first term	5 is not the degree of the first term
1 is the degree of the second term	3 is not the degree of the second term
2 is the degree of the third term	−6 is not the degree of the third term
4 is the degree of the fourth term	−1 is not the degree of the fourth term

Cooperative Learning and Algebra 1: Becky Bride
Kagan Publishing • 1 (800) 933-2667 • www.KaganOnline.com

ACTIVITY

1 DEFINE ME

3. Degree of a polynomial

Examples	Counterexamples

$$4x^2 + 8x - 7$$

2 is the degree of the polynomial   4, 8, 1, and −7 are not the degree

$$x^6 + 2x^5 + 3x - 4$$

6 is the degree of the polynomial   1, 2, 5, 3, and −4 are not the degree

$$5 + 3x - 6x^2 - x^4$$

4 is the degree of the polynomial   5, 3, −6, 2, 1, and −1 are not the degree

4. Coefficient of each term

Examples	Counterexamples

$$4x^2 + 8x - 7$$

4 is the coefficient of the first term   2 is not the coefficient of the first term

8 is the coefficient of the second term   1 is not the coefficient of the second term

−7 is the coefficient of the last term   0 is not the coefficient of the third term

$$x^6 + 2x^5 + 3x - 4$$

1 is the coefficient of the first term   6 is not the coefficient of the first term

2 is the coefficient of the second term   5 is not the coefficient of the second term

3 is the coefficient of the third term   1 is not the coefficient of the third term

−4 is the coefficient of the fourth term   0 is not the coefficient of the fourth term

$$5 + 3x - 6x^2 - x^4$$

5 is the coefficient of the first term   0 is not the coefficient of the first term

3 is the coefficient of the second term   1 is not the coefficient of the second term

−6 is the coefficient of the third term   2 is not the coefficient of the third term

−1 is the coefficient of the fourth term   4 is not the coefficient of the fourth term

Cooperative Learning and Algebra 1: Becky Bride
Kagan Publishing • 1 (800) 933-2667 • www.KaganOnline.com

ACTIVITY

1 DEFINE ME

5. Leading coefficient

Examples	Counterexamples
	$4x^2 + 8x - 7$
4 is the leading coefficient	2, 8, 1, and −7 are not the leading coefficient
	$2x^5 - x^6 + 3x - 4$
−1 is the leading coefficient	2, 5, 3, 6, 1 and −4 are not the leading coefficient
	$5 + 3x - 6x^2 - x^4$
−1 is the leading coefficient	5, 3, -6, 2, and 4 are not the leading coefficient

6. Standard form of a polynomial

Examples	Counterexamples
$4x^2 - 5x + 1$	$4x^2 - 5x^3 + 1$
$6x^3 + 7x^2 - 3$	$7x^2 + 6x^3 - 3$
$8x^4 - 5x^2 + x - 6$	$-5x^2 + x - 6 + 8x^4$
$x^9 - 3x^7 + 2x^6 - 5x + 8$	$-3x^7 + 2x^6 - 5x + x^9 + 8$

7. Opposite of a polynomial

Examples	Counterexamples
	$4x^2 + 8x - 7$
$-4x^2 - 8x + 7$ is the opposite	$-4x^2 + 8x - 7$ is not the opposite
	$2x^5 - x^6 + 3x - 4$
$-2x^5 + x^6 - 3x + 4$ is the opposite	$-2x^5 - x^6 - 3x - 4$ is not the opposite
	$5 + 3x - 6x^2 - x^4$
$-5 - 3x + 6x^2 - x^4$ is the opposite	$-5 + 3x - 6x^2 + x^4$ is not the opposite

Cooperative Learning and Algebra 1: Becky Bride
Kagan Publishing • 1 (800) 933-2667 • www.KaganOnline.com

2-8 VOCABULARY PRACTICE

Structure: RallyRobin

1. $4x^5 - 3x + 6x^2 - x^3 + 1$

2. $x - 3x^4 + 7x^8 - 2x^2 + 4$

3. $5 + b - 2b^2$

4. $7r - 8 - 3r^4 + 5r^2 + 2r^3$

5. $4 - 3h - 2h^3$

6. $x^2 - 7x^3 - x^5 - 2$

9 NAME MY TYPE

Structure: RallyRobin

State the type of each polynomial.

1. $3x^2 - 5x + 4$

2. $6h$

3. $9 - k$

4. $7 - m^2 + 5m$

5. $1 - 3x^2$

6. $9p^3 - 5p + 2$

Answers:
1. trinomial
2. monomial
3. binomial
4. trinomial
5. binomial
6. trinomial

Cooperative Learning and Algebra 1: Becky Bride
Kagan Publishing • 1 (800) 933-2667 • www.KaganOnline.com

LESSON 2
OPERATIONS ON POLYNOMIALS

This lesson contains activities to process operations on polynomials. The division activity has monomial divisors. The application activity involves area and perimeter, and some of the problems require students to translate words into mathematical symbols, reinforcing previous concepts. This activity also makes a connection between algebra and geometry.

1 ADD ME

Algebraic

Setup:
In pairs, Student A is the Sage; Student B is the Scribe. Students fold a sheet of paper in half and each writes his/her name on one half.

1. The Sage tells the Scribe how to add the polynomials for problem 1.

2. The Scribe records the Sage's instructions step-by-step in writing.

3. If the Sage is correct, the Scribe praises the Sage. Otherwise, the Scribe coaches, then praises.

4. Students switch roles for the next problem.

▶ **Structure**
• Sage-N-Scribe

▶ **Materials**
• Transparency 6.2.1
• 1 sheet of paper and pencil per pair

ACTIVITY
2 SUBTRACT ME

Algebraic

Setup:
In pairs, Student A is the Sage; Student B is the Scribe. Students fold a sheet of paper in half and each writes his/her name on one half.

1. The Sage tells the Scribe how to subtract the polynomials for problem 1.

2. The Scribe records the Sage's instructions step-by-step in writing.

3. If the Sage is correct, the Scribe praises the Sage. Otherwise, the Scribe coaches, then praises.

4. Students switch roles for the next problem.

▶ **Structure**
• Sage-N-Scribe

▶ **Materials**
• Transparency 6.2.2
• 1 sheet of paper and pencil per pair

ACTIVITY

3 MULTIPLY ME

▶ **Structure**
• Sage-N-Scribe

▶ **Materials**
• Transparency 6.2.3
• 1 sheet of paper and pencil per pair

Algebraic

Setup:
In pairs, Student A is the Sage; Student B is the Scribe. Students fold a sheet of paper in half and each writes his/her name on one half.

1. The Sage tells the Scribe how to multiply the binomials for problem 1.

2. The Scribe records the Sage's instructions step-by-step in writing.

3. If the Sage is correct, the Scribe praises the Sage. Otherwise, the Scribe coaches, then praises.

4. Students switch roles for the next problem.

ACTIVITY

4 MULTIPLY MY BINOMIALS

▶ **Structure**
• Sage-N-Scribe

▶ **Materials**
• Transparency 6.2.4
• 1 sheet of paper and pencil per pair

Algebraic

Setup:
In pairs, Student A is the Sage; Student B is the Scribe. Students fold a sheet of paper in half and each writes his/her name on one half.

1. The Sage tells the Scribe how to multiply the binomials for problem 1.

2. The Scribe records the Sage's instructions step-by-step in writing.

3. If the Sage is correct, the Scribe praises the Sage. Otherwise, the Scribe coaches, then praises.

4. Students switch roles for the next problem.

ACTIVITY

5 MULTIPLY MY POLYNOMIALS

▶ **Structure**
• Sage-N-Scribe

▶ **Materials**
• Transparency 6.2.5
• 1 sheet of paper and pencil per pair

Algebraic

Setup:
In pairs, Student A is the Sage; Student B is the Scribe. Students fold a sheet of paper in half and each writes his/her name on one half.

1. The Sage tells the Scribe how to multiply the polynomials for problem 1.

2. The Scribe records the Sage's instructions step-by-step in writing.

3. If the Sage is correct, the Scribe praises the Sage. Otherwise, the Scribe coaches, then praises.

4. Students switch roles for the next problem.

6 APPLY ME

Algebraic

Setup:
In pairs, Student A is the Sage; Student B is the Scribe. Students fold a sheet of paper in half and each writes his/her name on one half.

1. The Sage tells the Scribe how to solve problem 1.

2. The Scribe records the Sage's instructions step-by-step in writing.

3. If the Sage is correct, the Scribe praises the Sage. Otherwise, the Scribe coaches, then praises.

4. Students switch roles for the next problem.

▶ **Structure**
• Sage-N-Scribe

▶ **Materials**
• 1 Blackline 6.2.6 per pair
• 1 sheet of paper and pencil per pair

7 SIMPLIFY MY DIVISION

Algebraic

Setup:
In pairs, Student A is the Sage; Student B is the Scribe. Students fold a sheet of paper in half and each writes his/her name on one half.

1. The Sage tells the Scribe how to simplify problem 1.

2. The Scribe records the Sage's instructions step-by-step in writing.

3. If the Sage is correct, the Scribe praises the Sage. Otherwise, the Scribe coaches, then praises.

4. Students switch roles for the next problem.

▶ **Structure**
• Sage-N-Scribe

▶ **Materials**
• Transparency 6.2.7
• 1 sheet of paper and pencil per pair

ACTIVITY

8

EXPLORING COMPLETING THE SQUARE

▶ **Structure**
• Solo-Pair Consensus-Team Consensus

▶ **Materials**
• 1 Blackline 6.2.8 per student
• 1 sheet of paper and pen/pencil per student
• 1 set of algebra tiles (see chapter 1)
• 1 highlighter per student

This activity is included because in Chapter 8, Lesson 2, students will rewrite a quadratic function given in standard form into vertex form. To do this they must be able to complete the square.

Exploratory

Solo

1. Have each student complete the investigation individually.

Pair Consensus

2. For each problem on the investigation, each student shares with his/her partner, using RallyRobin, his/her response. They discuss the

problems on which they disagree on, trying to come to consensus on the correct response. For the problems they can't reach consensus on, they mark these so they can focus on them during the team phase. Encourage the students to add to their responses if their partner verbalized an understanding they did not see.

Team Consensus

3. Each pair shares their responses, using RallyRobin, with the other pair in their team, augmenting their responses if necessary. When the teams are through sharing, each student should have a detailed and complete summary.

ACTIVITY

9

COMPLETE MY SQUARE

▶ **Structure**
• Sage-N-Scribe

▶ **Materials**
• Transparency 6.2.9
• 1 sheet of paper and pencil per pair

Algebraic

Setup:
In pairs, Student A is the Sage; Student B is the Scribe. Students fold a sheet of paper in half and each writes his/her name on one half.

1. The Sage tells the Scribe how to complete the square for problem 1.

2. The Scribe records the Sage's instructions step-by-step in writing.

3. If the Sage is correct, the Scribe praises the Sage. Otherwise, the Scribe coaches, then praises.

4. Students switch roles for the next problem.

Structure: Sage-N-Scribe

Add the following polynomials. Answers need to be in standard form.

1. $(4x^2 - 3x + 1) + (9x - 3)$

2. $(6x - 2x^4 + 3x^2) + (5x^4 - x^3 + 7x - x^2)$

3. $(1.2h + 3.5h^2) + (4.5h^2 - 6h + 3)$

4. $(4x - 3x^4 - 2x^3 + 3) + (8 - x^4 + 2x^3 + x^2)$

Answers:
1. $4x^2 + 6x - 2$
2. $3x^4 - x^3 + 2x^2 + 13x$
3. $8h^2 - 4.8h + 3$
4. $-4x^4 + x^2 + 4x + 11$

2 SUBTRACT ME

Structure: Sage-N-Scribe

Subtract the following polynomials. Answers need to be in standard form.

1. $(4x^2 - 3x + 1) - (9x - 3)$

2. $(6x - 2x^4 + 3x^2) - (5x^4 - x^3 + 7x - x^2)$

3. $(1.2h + 3.5h^2) - (4.5h^2 - 6h + 3)$

4. $(4x - 3x^4 - 2x^3 + 3) - (8 - x^4 + 2x^3 + x^2)$

Answers:
1. $4x^2 - 12x + 4$
2. $-7x^4 + x^3 + 4x^2 - x$
3. $-1h^2 + 7.2h - 3$
4. $-2x^4 - 4x^3 - x^2 + 4x - 5$

Cooperative Learning and Algebra 1: Becky Bride
Kagan Publishing • 1 (800) 933-2667 • www.KaganOnline.com

MULTIPLY ME

Structure: Sage-N-Scribe

Multiply and write the answer in standard form.

1. $-3b^3(5 + b - 2b^2)$

2. $4h(4 - 3h - 2h^3)$

3. $5x^4(x - 3x^4 + 7x^8 - 2x^2 + 4)$

4. $-2r^2(7r - 8 - 3r^4 + 5r^2 + 2r^3)$

Answers:
1. $6b^5 - 3b^4 - 15b^3$
2. $-8h^4 - 12h^2 + 16h$
3. $35x^{12} - 15x^8 - 10x^6 + 5x^5 + 20x^4$
4. $6r^6 - 4r^5 - 10r^4 + 14r^3 + 16r^2$

Structure: Sage-N-Scribe

Multiply and write the answer in standard form.

1. $(2x - 7)(5x + 1)$

2. $(3x - 8)(x + 1)$

3. $(5x + 1)(2x - 3)$

4. $(6x + 5)(4 - 3x)$

5. $(7 - 2x)(8 + 3x)$

6. $(9 - 2x)(3x + 4)$

Answers:
1. $10x^2 - 33x - 7$
2. $3x^2 - 5x - 8$
3. $10x^2 - 13x - 3$
4. $-18x^2 + 9x + 20$
5. $-6x^2 + 5x + 56$
6. $-6x^2 + 19x + 36$

Cooperative Learning and Algebra 1: Becky Bride
Kagan Publishing • 1 (800) 933-2667 • www.KaganOnline.com

MULTIPLY MY POLYNOMIALS

Structure: Sage-N-Scribe

Multiply and write the answer in standard form.

1. $(2x - 7)(5x^2 - x + 1)$

2. $(3x - 8)(x^2 - 4x + 1)$

3. $(5h + 1)(2h^2 + 6h - 3)$

4. $(6m + 5)(4 - 5m^2 + 3m)$

5. $(7 - 2x)(8 + 3x - x^2)$

6. $(1 - 2x)(3x^2 - 2x + 4)$

Answers:
1. $10x^3 - 37x^2 + 9x - 7$
2. $3x^3 - 20x^2 + 35x - 8$
3. $10h^3 + 32h^2 - 9h - 3$
4. $-30m^3 - 7m^2 + 39m + 20$
5. $-2x^3 - 13x^2 + 5x + 56$
6. $-6x^3 + 7x^2 - 10x + 4$

ACTIVITY

6 APPLY ME

Structure: Sage-N-Scribe

1. If the length of a rectangle is two more than three times its width, then write an expression for the area of the rectangle.

2. Find the perimeter of the rectangle in problem 1.

3. One side of a triangular flag is two more than five times the third side. The second side of the flag is seven less than the third side. Find an expression for the perimeter of the flag.

4. The base of a triangular sail is four less than twice some number, and its height is one more than that same number. Find an expression for the area of the sail.

5. The radius of a circular tablecloth is nine more than four times a number. Find an expression for the circumference of the tablecloth.

6. Find an expression for the area of the tablecloth in problem 5.

7. Find an expression for the area of the figure below.

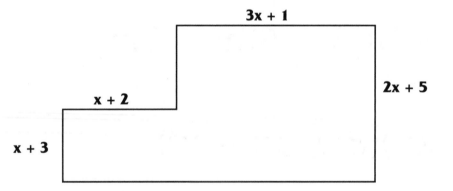

8. Find an expression for the perimeter of the figure above.

Answers:
1. $2w + 3w^2$
2. $4 + 8w$
3. $7x - 5$
4. $2x^2 - 2x - 4$
5. $\pi(8x + 18)$
6. $(16x^2 + 72x + 81)\pi$
7. $7x^2 + 22x + 11$
8. $12x + 16$

Cooperative Learning and Algebra 1: Becky Bride
Kagan Publishing • 1 (800) 933-2667 • www.KaganOnline.com

ACTIVITY
7 SIMPLIFY MY DIVISION

POLYNOMIALS
5xy

Structure: Sage-N-Scribe

Simplify each expression.

1. $\dfrac{6x^3 + 4x^2 - 10x}{4x}$

2. $\dfrac{12x^4 + 14x^2 - 18x^3}{2x^2}$

3. $\dfrac{15m^4 + 6m^5 - 18m^3}{3m^3}$

4. $\dfrac{-15m^4 - 5m^5}{-5m^3}$

5. $\dfrac{-24p^5 + 36p^8}{-12p^4}$

6. $\dfrac{10xw^4 + 6xw^3 - 18x^3w^2}{8x}$

Answers:
1. $\frac{3}{2}x^2 + x - \frac{5}{2}$
2. $6x^2 - 9x + 7$
3. $2m^2 + 5m - 6$
4. $m^2 + 3m$
5. $-3p^4 + 2p$
6. $\frac{5}{4}w^4 + \frac{3}{4}w^3 - \frac{9}{4}x^2w^2$

Cooperative Learning and Algebra 1: Becky Bride
Kagan Publishing • 1 (800) 933-2667 • www.KaganOnline.com

ACTIVITY 8 EXPLORING COMPLETING THE SQUARE

1. Place algebra tiles on your workspace that model the binomial $x^2 + 2x$.

2. Arrange the tiles chosen in problem 1 so that they begin to form a square. You will be missing one corner of the square.

3. How many tiles that represent 1 unit must be added to your tiles to **complete** the square?

4. On your sheet of paper, draw a picture of the square you made with algebra tiles, highlighting the unit tiles you had to add. Write the new polynomial that is represented by the tiles in your square. It should now be a trinomial.

5. Place algebra tiles on your workspace that model the binomial $x^2 + 4x$.

6. Arrange the tiles chosen in problem 5 so that they begin to form a square. You will be missing one corner of the square.

7. How many tiles that represent 1 unit must be added to your tiles to **complete** the square?

8. On your sheet of paper, draw a picture of the square you made with algebra tiles, highlighting the unit tiles you had to add. Write the new polynomial that is represented by the tiles in your square. It should now be a trinomial.

9. Place algebra tiles on your workspace that model the binomial $x^2 + 10x$.

10. Arrange the tiles chosen in problem 9 so that they begin to form a square. You will be missing one corner of the square.

11. How many tiles that represent 1 unit must be added to your tiles to **complete** the square?

Cooperative Learning and Algebra 1: Becky Bride
Kagan Publishing • 1 (800) 933-2667 • www.KaganOnline.com

EXPLORING COMPLETING THE SQUARE

12. On your sheet of paper, draw a picture of the square you made with algebra tiles, highlighting the unit tiles you had to add. Write the new polynomial that is represented by the tiles in your square. It should now be a trinomial.

13. Place algebra tiles on your workspace that model the binomial $x^2 - 8x$.

14. Arrange the tiles chosen in problem 13 so that they begin to form a square. You will be missing one corner of the square.

15. How many tiles that represent 1 unit must be added to your tiles to **complete** the square?

16. On your sheet of paper, draw a picture of the square you made with algebra tiles, highlighting the unit tiles you had to add. Write the new polynomial that is represented by the tiles in your square. It should now be a trinomial.

17. Place algebra tiles on your workspace that model the binomial $x^2 - 12x$.

18. Arrange the tiles chosen in problem 17 so that they begin to form a square. You will be missing one corner of the square.

19. How many tiles that represent 1 unit must be added to your tiles to **complete** the square?

20. On your sheet of paper, draw a picture of the square you made with algebra tiles, highlighting the unit tiles you had to add. Write the new polynomial that is represented by the tiles in your square. It should now be a trinomial.

21. Why is the process you just completed five times called **completing** the square?

EXPLORING COMPLETING THE SQUARE

22. In order to begin completing the square each time, what did you have to do to the x tiles?

23. Complete the following: How many unit tiles were added in
    a) problem 3
    b) problem 7
    c) problem 11
    d) problem 15
    e) problem 19
    to complete the square?

24. Look at the answers to a-e of problem 13.
    a) What is special about each one of these numbers?
    b) How are these numbers related to the coefficient of each x term in each problem?

25. Based on your answers to problems 22-24, how many tiles have to be added to $x^2 + 20x$ to complete the square?

26. Your best friend was absent today. Write your friend a letter explaining how to complete the square using algebra tiles and how to do it without using algebra tiles.

Cooperative Learning and Algebra 1: Becky Bride
Kagan Publishing • 1 (800) 933-2667 • www.KaganOnline.com

9 COMPLETE MY SQUARE

Structure: Sage-N-Scribe

a) For each binomial, what must be added to complete the square?
b) Write the trinomial formed when the square is completed.

1. $x^2 - 14x$

2. $x^2 + 18x$

3. $x^2 + 3x$

4. $x^2 - 5x$

Answers:
1. a) 49; b) $x^2 - 14x + 49$
2. a) 81; b) $x^2 + 18x + 81$
3. a) $\frac{9}{4}$; b) $x^2 + 3x + \frac{9}{4}$
4. a) $\frac{25}{4}$; b) $x^2 - 5x + \frac{25}{4}$

LESSON 3
FACTORING AND SOLVING POLYNOMIAL EQUATIONS

This lesson's focus is on factoring. Trinomial factoring and difference of two squares factoring are explored by students by comparing and contrasting two binomial factors to their respective trinomial. Activities to process factoring are included. The remainder of the lesson then uses factoring to solve quadratic equations. The lesson wraps up with applications and an activity to synthesize what they have learned in this lesson.

1 WHAT IS THERE IN COMMON?

▶ **Structure**
· RallyRobin

▶ **Materials**
· Transparency 6.3.1

This activity prepares the students for the next activity. All that is required is that the students can determine the common factor. This skill must be in place before students can be successful with Activity 2.

Algebraic

1. Teacher poses multiple problems with Transparency 6.3.1.

2. In pairs, students take turns stating the common factor of the polynomial, orally.

ACTIVITY

2 COMMON FACTOR

▶ **Structure**
· Sage-N-Scribe

▶ **Materials**
· Transparency 6.3.1
· 1 sheet of paper and pencil per pair

Algebraic

Setup:
In pairs, Student A is the Sage; Student B is the Scribe. Students fold a sheet of paper in half and each writes his/her name on one half.

1. The Sage tells the Scribe how to factor problem 1.

2. The Scribe records the Sage's instructions step-by-step in writing.

3. If the Sage is correct, the Scribe praises the Sage. Otherwise, the Scribe coaches, then praises.

4. Students switch roles for the next problem.

Cooperative Learning and Algebra 1: Becky Bride
Kagan Publishing • 1 (800) 933-2667 • www.KaganOnline.com

3 EXPLORING TRINOMIAL FACTORING

This activity prepares students to factor trinomials with the trial and error method. If you use another method, then this activity should be skipped.

This activity has students examine the multiplication of binomials to understand how to factor a trinomial into two binomials. The expressions in all four pairs of binomials are the same. The only difference is the operation signs in each binomial. The first goal of this investigation is for students to see that the first terms of the binomial factors must be factors of the first term of the trinomial. These problems were written with leading coefficients not equal to 1. When factoring is taught first with leading coefficients of 1, the students have a difficult time generalizing this to leading coefficients that are not 1. Leading coefficients of 1 are just a special case of factoring trinomials. By beginning with leading coefficients that are not 1, the students are tuned into considering the factors of the leading coefficient, not just the factors of the last term. This is an activity that works best with processing done throughout the activity. After problem 10 on the exploration, the trans-

parency for Activity 4 can be used to process the possible factors for the first term of the binomials. This helps students digest the exploration one small piece at a time, which enhances learning.

Process again after problem 13 using Transparency 6.3.3 by having students list the possible last terms of the binomials, based on the factors of the last term of the trinomial. After problem 17, process the operation symbols that should be used in the binomial factors using Transparency 6.3.3. The remainder of the investigation uses the information that the students examined and demonstrates the factoring of several trinomials.

Exploratory

Solo

1. Have each student complete the investigation individually.

Pair Consensus

2. For each problem on the investigation, each student shares with his/her partner, using RallyRobin, his/her

▶ **Structure**
· Solo-Pair Consensus-Team Consensus

▶ **Materials**
· 1 Blackline 6.3.2 per student
· 1 sheet of paper and pen/pencil per student

response. They discuss the problems on which they disagree on, trying to come to consensus on the correct response. For the problems they can't reach consensus on, they mark these so they can focus on them during the team phase. Encourage the students to add to their responses if their partner verbalized an understanding they did not see.

Team Consensus

3. Each pair shares their responses, using RallyRobin, with the other pair in their team, augmenting their responses if necessary. When the teams are through sharing, each student should have a detailed, complete summary.

ACTIVITY 4

FACTOR MY TRINOMIAL

▶ **Structure**
• Sage-N-Scribe

▶ **Materials**
• Transparency 6.3.3
• 1 sheet of paper and pencil per pair

Algebraic

Setup:
In pairs, Student A is the Sage; Student B is the Scribe. Students fold a sheet of paper in half and each writes his/her name on one half.

1. The Sage tells the Scribe how to factor the polynomial for problem 1.

2. The Scribe records the Sage's instructions step-by-step in writing.

3. If the Sage is correct, the Scribe praises the Sage. Otherwise, the Scribe coaches, then praises.

4. Students switch roles for the next problem.

ACTIVITY 5

EXPLORING THE DIFFERENCE OF TWO SQUARES

▶ **Structure**
• Solo-Pair Consensus-Team Consensus

▶ **Materials**
• 1 Blackline 6.3.4 per student
• 1 sheet of paper and pen/ pencil per student

Exploratory

Solo

1. Have each student complete the investigation individually.

Pair Consensus

2. For each problem on the investigation, each student shares with his/her partner, using RallyRobin, his/her response. They discuss the problems on which they disagree on, trying to come to consensus on the correct response. For the problems

they can't reach consensus on, they mark these so they can focus on them during the team phase. Encourage the students to add to their responses if their partner verbalized an understanding they did not see.

Team Consensus

3. Each pair compares their responses to the responses of the other pair in their team, augmenting their responses if necessary. When the teams are through sharing, each student should have a detailed, complete summary.

Cooperative Learning and Algebra 1: Becky Bride
Kagan Publishing • 1 (800) 933-2667 • www.KaganOnline.com

ACTIVITY 6 — FACTORING THE DIFFERENCE OF TWO SQUARES

Algebraic

Setup:
In pairs, Student A is the Sage; Student B is the Scribe. Students fold a sheet of paper in half and each writes his/her name on one half.

1. The Sage tells the Scribe how to factor the polynomial for problem 1.

2. The Scribe records the Sage's instructions step-by-step in writing.

3. If the Sage is correct, the Scribe praises the Sage. Otherwise, the Scribe coaches, then praises.

4. Students switch roles for the next problem.

▶ **Structure**
• Sage-N-Scribe

▶ **Materials**
• Transparency 6.3.5
• 1 sheet of paper and pencil per pair

ACTIVITY 7 — FACTOR ME COMPLETELY

Algebraic

Setup:
In pairs, Student A is the Sage; Student B is the Scribe. Students fold a sheet of paper in half and each writes his/her name on one half.

1. The Sage tells the Scribe how to completely factor the polynomial for problem 1.

2. The Scribe records the Sage's instructions step-by-step in writing.

3. If the Sage is correct, the Scribe praises the Sage. Otherwise, the Scribe coaches, then praises.

4. Students switch roles for the next problem.

▶ **Structure**
• Sage-N-Scribe

▶ **Materials**
• Transparency 6.3.6
• 1 sheet of paper and pencil per pair

ACTIVITY 8 — FIND MY SOLUTION

Algebraic

Setup:
In pairs, Student A is the Sage; Student B is the Scribe. Students fold a sheet of paper in half and each writes his/her name on one half.

1. The Sage tells the Scribe how to solve problem 1.

2. The Scribe records the Sage's instructions step-by-step in writing.

3. If the Sage is correct, the Scribe praises the Sage. Otherwise, the Scribe coaches, then praises.

4. Students switch roles for the next problem.

▶ **Structure**
• Sage-N-Scribe

▶ **Materials**
• Transparency 6.3.7
• 1 sheet of paper and pencil per pair

Chapter 6: Polynomials
Lesson Three

ACTIVITY

9 SOLVE ME

▶ **Structure**
• Sage-N-Scribe

▶ **Materials**
• Transparency 6.3.8
• 1 sheet of paper and pencil per pair

Algebraic

Setup:
In pairs, Student A is the Sage; Student B is the Scribe. Students fold a sheet of paper in half and each writes his/her name on one half.

1. The Sage tells the Scribe how to solve problem 1.

2. The Scribe records the Sage's instructions step-by-step in writing.

3. If the Sage is correct, the Scribe praises the Sage. Otherwise, the Scribe coaches, then praises.

4. Students switch roles for the next problem.

ACTIVITY

10 POLYNOMIAL APPLICATIONS

▶ **Structure**
• Sage-N-Scribe

▶ **Materials**
• Transparency 6.3.9
• 1 sheet of paper and pencil per pair

Algebraic

Setup:
In pairs, Student A is the Sage; Student B is the Scribe. Students fold a sheet of paper in half and each writes his/her name on one half.

1. The Sage tells the Scribe how to solve problem 1.

2. The Scribe records the Sage's instructions step-by-step in writing.

3. If the Sage is correct, the Scribe praises the Sage. Otherwise, the Scribe coaches, then praises.

4. Students switch roles for the next problem.

Cooperative Learning and Algebra 1: Becky Bride
Kagan Publishing • 1 (800) 933-2667 • www.KaganOnline.com

WHAT HAVE WE LEARNED?

Synthesis via Graphic Organizer

1. Each teammate signs his/her name in the upper-right corner of the team paper with the color pen/pencil he/she is using.

2. One teammate writes "Polynomials" in the center of the team paper in a rectangle.

3. Teammate 1 shares with the team one core concept he/she learned in the unit.

4. The student checks for consensus.

5. The teammates show agreement or lack of agreement with thumbs up or down.

6. If there is agreement, the students celebrate and the teammate records the core concept on the graphic organizer connecting it with a line to the main idea, "Polynomials." If not, teammates discuss the response until there is agreement and then they celebrate.

7. Play continues with the next student's core concept until all core concepts are exhausted.

8. Repeat steps 3-7, with teammates adding details to each core concept and making bridges between related ideas.

▶ **Structure**
• RoundTable Consensus

▶ **Materials**
• 1 large sheet of paper per team
• different color pen or pencil for each student on the team

WHAT IS THERE IN COMMON?
COMMON FACTOR

ACTIVITIES 1 & 2

POLYNOMIALS 5xy

Structure: RallyRobin & Sage-N-Scribe

1. $4x^3 - 6x^2 + 10x$

2. $9y^4 - 12y^3 - 15y^2$

3. $18m^5 + 12m^3 - 6m^2$

4. $25wp + 30w^2p^3 - 15w^3p^4$

5. $14p^3n - 10p^2n^2 + 22pn^3$

6. $7d^6x^7 - 14d^5x^6 - 21d^4x^5$

Answers:
1. $2x(2x^2 - 3x + 5)$
2. $3y^2(3y^2 - 4y - 5)$
3. $6m^2(3m^3 + 2m - 1)$
4. $5wp(5 + 6wp^2 - 3w^2p^3)$
5. $2pn(7p^2 - 5pn + 11n^2)$
6. $7d^4x^5(d^2x^2 - 2dx - 3)$

Cooperative Learning and Algebra 1: Becky Bride
Kagan Publishing • 1 (800) 933-2667 • www.KaganOnline.com

EXPLORING TRINOMIAL FACTORING

Structure: Solo-Pair Consensus-Team Consensus

The goal of this investigation is for you to thoroughly examine the product of two binomials so that when you are given a trinomial like $6x^2 - 13x - 5$, you can factor it into its binomial factors of $(3x + 1)(2x - 5)$. In order to do this, you must examine the multiplication process in detail.

1. $(2x + 3)(4x + 5)$

2. $(2x - 3)(4x - 5)$

3. $(2x + 3)(4x - 5)$

4. $(2x - 3)(4x + 5)$

5. Look at each pair of binomials above. What is similar about all of them?

6. Look at each pair of binomials above. What is different about all of them?

ACTIVITY 3 — EXPLORING TRINOMIAL FACTORING

7. Multiply each pair of binomials in problems 1-4; showing all work in the space provided on page 353.

8. Look at the trinomial answers for problems 1-4 and the binomial factors. Why is the first term always $8x^2$?

9. Look at the trinomial in the description of the goal of this investigation at the top of page 353. The first term is $6x^2$ and the first terms of the binomial factors are $2x$ and $3x$. Explain why this is consistent with your explanation in problem 8.

10. Using the trinomial $12x^2 + 5x - 2$, write all the possible factors for the first terms of the binomial factors.

11. Look at the trinomial answers for problems 1-4. Why is the absolute value of the last term always 15?

12. Look at the trinomial in the description of the goal of this investigation at the top of page 353. The last term is 5 and the last terms of the binomial factors are 1 and 5. Explain why this is consistent with your explanation in problem 11.

Cooperative Learning and Algebra 1: Becky Bride
Kagan Publishing • 1 (800) 933-2667 • www.KaganOnline.com

EXPLORING TRINOMIAL FACTORING

13. Using the trinomial $12x^2 + 5x - 2$, write all the possible factors for the last terms of the binomial factors.

14. Look at the trinomial answers again for problems 1-4. Now focus on the operation symbol in front of the last term and the operation symbols in the binomial factors. Why are some of the last operation symbols in the trinomial addition and others subtraction?

15. Look at the trinomial in the description of the goal of this investigation at the top of page 353. The last operation symbol is subtraction and the last binomial factors have one addition sign and one subtraction sign. Explain why this is consistent with your explanation in problem 14.

16. Look at the trinomials in problems 1 and 2. The operation symbol in front of the last term is addition. Look at the operation symbols in the binomials and the operation symbol in front of the middle term. How will the operation symbol of the middle term tell you which operation symbols to use when the last operation symbol is addition?

17. Using the trinomial $12x^2 + 5x - 2$, what are the operation symbols for the binomial factors?

ACTIVITY 3 EXPLORING TRINOMIAL FACTORING

18. You have all the information to factor a trinomial. From this point on, you need to try different factors to see which combination gives you the correct middle term. The middle term is the sum of the product of the outside term and the inside terms of the binomial. Drawing the diagram below will be very helpful when determining which combination forms the correct middle term. Look at the example below.

Example 1

$2x^2 + 7x + 3$

The possible factors for the first term are $2x$ and x.
The possible factors for the last term are 3 and 1.
The operation symbols for the binomials are both $+$ because 3 is positive and the 7 is also positive.

The first possible combination is

$$2x$$
$$(2x + 3)\ (x + 1)$$
$$3x$$

but notice when $2x$ and $3x$ are added, the result is $5x$, which is not the middle term. Switch the 3 and the 1 and see what the new result will be.

$$6x$$
$$(2x + 1)\ (x + 3)$$
$$1x$$

notice when $6x$ and $1x$ are added, the result is $7x$, which is the middle term. So $2x^2 + 7x + 3$ factors into $(2x + 1)(x + 3)$.

Cooperative Learning and Algebra 1: Becky Bride
Kagan Publishing • 1 (800) 933-2667 • www.KaganOnline.com

EXPLORING TRINOMIAL FACTORING

Example 2

$10x^2 + x - 3$

The possible factors for the first term are 10x and 1x or 5x and 2x.
The possible factors for the last term are 1 and 3.
The operation symbols for the binomials are + and − because the last term has a subtraction symbol in front of it.

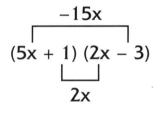

$(5x + 1) (2x - 3)$

Notice when −15x is combined with 2x, the result is −13x, which is not the middle term. Before changing factors, switch the 1 and the 3 because the combination will result in a different answer.

$(5x + 3) (2x - 1)$

Notice when −5x is combined with 6x, the result is 1x, which is the middle term. If the result had been a −1x, switching the + and − symbols would have fixed it.

4 **FACTOR MY TRINOMIAL**

Structure: Sage-N-Scribe

Factor each trinomial.

1. $3x^2 - 10x - 8$

2. $8m^2 - 14m - 15$

3. $6w^2 - 13w + 6$

4. $20y^2 - 7y - 6$

5. $4m^2 + 8m + 3$

6. $12x^2 + 7x - 10$

Answers:
1. $(3x + 2)(x - 4)$
2. $(4m + 3)(2m - 5)$
3. $(3w - 2)(2w - 3)$
4. $(4y - 3)(5y + 2)$
5. $(2m + 3)(2m + 1)$
6. $(4x + 5)(3x - 2)$

Cooperative Learning and Algebra 1: Becky Bride
Kagan Publishing • 1 (800) 933-2667 • www.KaganOnline.com

POLYNOMIALS
5xy

ACTIVITY 5
EXPLORING THE DIFFERENCE OF TWO SQUARES

Structure: Solo-Pair Consensus-Team Consensus

The purpose of this investigation is for you to find patterns when special binomials are multiplied that can be used to factor the result.

1. $(2x + 3)(2x - 3)$

2. $(3x + 4)(3x - 4)$

3. $(5x + 1)(5x - 1)$

4. $(4x + 3)(4x - 3)$

5. Look at the binomial factors above. Describe how they are similar.

6. Look at the binomial factors above. Describe how they are different.

EXPLORING THE DIFFERENCE OF TWO SQUARES

7. Multiply each binomial pair in problems 1-4 in the space provided on page 359, showing all work.

8. Was the answer for each problem a binomial or trinomial?

9. Explain what happened to the middle terms in each problem and why it happened.

10. Look at the coefficient of the first terms in each answer. What is very special about these numbers?

11. Look at the coefficient of the last term in each answer. What is very special about these coefficients?

12. Based on your observations above, factor the following binomials.

a) $25x^2 - 16$

b) $49x^2 - 1$

c) $16x^2 - 25$

d) $81x^2 - 100$

Cooperative Learning and Algebra 1: Becky Bride
Kagan Publishing • 1 (800) 933-2667 • www.KaganOnline.com

FACTORING THE DIFFERENCE OF TWO SQUARES

Structure: Sage-N-Scribe

Factor each of the following. If it is not factorable, write "prime."

1. $x^2 - 49$

2. $4x^2 - 81$

3. $5y^2 - 20$

4. $64 - 9m^2$

5. $x^2 + 4$

6. $8x^2 - 18$

Answers:
1. $(x + 7)(x - 7)$
2. $(2x + 9)(2x - 9)$
3. $5(y + 2)(y - 2)$
4. $(8 + 3m)(8 - 3m)$
5. prime
6. $2(2x + 3)(2x - 3)$

FACTOR ME COMPLETELY

POLYNOMIALS 5Xy

Structure: Sage-N-Scribe

Factor each expression completely.

1. $8x^2 - 32$

2. $6x^3 + 15x^2 + 9x$

3. $3x^3 + 19x^2 + 20x$

4. $x^4 - 1$

5. $12x^4 + x^2 - 1$

6. $x^4 - 13x^2 + 36$

7. $8x^3 + 26x^2 - 7x$

Answers:
1. $8(x + 2)(x - 2)$
2. $3x(2x + 3)(x + 1)$
3. $x(3x + 4)(x + 5)$
4. $(x^2 + 1)(x + 1)(x - 1)$
5. $(2x + 1)(2x - 1)(3x^2 + 1)$
6. $(x + 2)(x - 2)(x + 3)(x - 3)$
7. $x(4x - 1)(2x + 7)$
8. $3(6x - 5)(x + 2)$

8. $18x^2 + 21x - 30$

Cooperative Learning and Algebra 1: Becky Bride
Kagan Publishing • 1 (800) 933-2667 • www.KaganOnline.com

ACTIVITY

8 **FIND MY SOLUTION**

Structure: Sage-N-Scribe

Find the solution(s) for each of the following.

1. $(3x + 1)(4x - 7) = 0$

2. $(4u - 3)(u + 1) = 0$

3. $m (4 - m)(5 + m) = 0$

4. $(-2w)(2w - 3)(3w + 10) = 0$

5. $(5p - 3)(4 - p)(8p + 1) = 0$

6. $d(2d - 1)(3d + 2)(8 - d) = 0$

Answers:

1. $\left\{\dfrac{-1}{3}, \dfrac{7}{4}\right\}$

2. $\left\{\dfrac{3}{4}, -1\right\}$

3. $\{-5, 0, 4\}$

4. $\left\{\dfrac{-10}{3}, 0, \dfrac{3}{2}\right\}$

5. $\left\{\dfrac{-1}{8}, \dfrac{3}{5}, 4\right\}$

6. $\left\{\dfrac{-2}{3}, 0, \dfrac{1}{2}, 8\right\}$

SOLVE ME

Structure: Sage-N-Scribe

Find the solution(s) for each of the following.

1. $-x^2 - 6x = 0$

2. $7x - 2x^2 = 0$

3. $4x^2 - 9 = 0$

4. $x^2 - 10x - 75 = 0$

5. $4w^2 + 5w = 6$

6. $2p^2 - p = 1$

7. $3u^3 + u = 4u^2$

8. $3h^2 - 48 = 0$

Answers:
1. $\{-6, 0\}$
2. $\left\{0, \dfrac{7}{2}\right\}$
3. $\left\{\dfrac{-3}{2}, \dfrac{3}{2}\right\}$
4. $\{-5, 15\}$
5. $\left\{-2, \dfrac{3}{4}\right\}$
6. $\left\{\dfrac{-1}{2}, 1\right\}$
7. $\left\{1, \dfrac{1}{3}, 0\right\}$
8. $\{-4, 4\}$

Cooperative Learning and Algebra 1: Becky Bride
Kagan Publishing • 1 (800) 933-2667 • www.KaganOnline.com

ACTIVITY
10 POLYNOMIAL APPLICATIONS

Structure: Sage-N-Scribe

For each problem below:
  a) **Draw a picture if applicable.**
  b) **Define a variable.**
  c) **Write an equation that models the problem.**
  d) **Find the solution.**

1. One side of a rectangular flower bed is two feet less than the other side. The area of the flower bed is 8 square feet. Find the dimensions of the flower bed.

2. The base of a triangle is six more than twice the height. If the area of the triangle is 10 square inches, find the length of the height.

3. The distance a bicyclist travels can be represented by $D = 8t^2 + 2t$, where D is the distance the cyclist travels, in feet, and t is the time in seconds for the cyclist to travel that distance. How long does it take the cyclist to travel 15 feet?

4. The distance a bunny travels can be represented by $D = 18t^2 + 9t$, where D is the distance the bunny travels, in feet, and t is the time in seconds for the bunny to travel that distance. How long does it take the bunny to travel 20 feet?

Answers:
1. 2 ft x 4 ft
2. 2 inches
3. 1.25 seconds
4. 0.83 seconds

RADICALS

This lesson works with square roots. Lesson 1 extends the sets of numbers defined in Chapter 1 to include irrational numbers. The remaining part of the lesson involves simplifying square roots. Lesson 2 has students perform operations involving square roots. Multiplying radicals reinforces the multiplication of polynomials studied in Chapter 6. The final lesson applies square roots and connects algebra to geometry.

LESSON 1 SIMPLIFYING RADICALS

ACTIVITY 1: Name My Sets
ACTIVITY 2: Simplify My Radical
ACTIVITY 3: Simplify My Radical Fraction
ACTIVITY 4: Simplify My Radical Fraction—Advanced

LESSON 2 OPERATIONS WITH RADICALS

ACTIVITY 1: Add/Subtract My Radicals
ACTIVITY 2: Add/Subtract My Radicals—Advanced
ACTIVITY 3: Multiply My Radicals
ACTIVITY 4: Distribute My Radical
ACTIVITY 5: Multiply My Radicals—Advanced

LESSON 3 APPLICATIONS

ACTIVITY 1: Area and Radicals
ACTIVITY 2: Pythagorean Theorem
ACTIVITY 3: Applications with the Pythagorean Theorem
ACTIVITY 4: What Did We Learn?

LESSON 1
SIMPLIFYING RADICALS

This lesson works with square roots only. This chapter is included because simplifying radicals is necessary in geometry (for the special right triangles) and trigonometry. This concept will be explored in more depth in Algebra 2 with roots other than square roots. Also complex numbers require the skills involved in simplifying radicals. Therefore, it is important that the students have a foundation on which to build.

1 NAME MY SETS

▶ **Structure**
• RallyCoach

▶ **Materials**
• Transparency 7.1.1
• 1 sheet of paper and pencil per pair

Algebraic

1. Teacher poses multiple problems with Transparency 7.1.1.

2. Partner A writes the set(s) of numbers to which the given number belongs.

3. Partner B watches and listens, checks, and praises.

4. Partner B writes the set(s) of numbers to which the number in the next problem belongs.

5. Partner A watches and listens, checks, and praises.

6. Repeat for remaining problems starting at step 2.

2 SIMPLIFY MY RADICAL

▶ **Structure**
• Sage-N-Scribe

▶ **Materials**
• Transparency 7.1.2
• 1 sheet of paper and pencil per pair

Algebraic

Setup:
In pairs, Student A is the Sage; Student B is the Scribe. Students fold a sheet of paper in half and each writes his/her name on one half.

1. The Sage tells the Scribe how to simplify the radical in problem 1.

2. The Scribe records the Sage's instructions step-by-step in writing.

3. If the Sage is correct, the Scribe praises the Sage. Otherwise, the Scribe coaches, then praises.

4. Students switch roles for the next problem.

Cooperative Learning and Algebra 1: Becky Bride
Kagan Publishing • 1 (800) 933-2667 • www.KaganOnline.com

ACTIVITY

3 SIMPLIFY MY RADICAL FRACTION

Numeric

Setup:
In pairs, Student A is the Sage; Student B is the Scribe. Students fold a sheet of paper in half and each writes his/her name on one half.

1. The Sage tells the Scribe how to simplify the radical in problem 1.

2. The Scribe records the Sage's instructions step-by-step in writing.

3. If the Sage is correct, the Scribe praises the Sage. Otherwise, the Scribe coaches, then praises.

4. Students switch roles for the next problem.

▶ **Structure**
• Sage-N-Scribe

▶ **Materials**
• Transparency 7.1.3
• 1 sheet of paper and pencil per pair

ACTIVITY

4 SIMPLIFY MY RADICAL FRACTION—ADVANCED

Numeric

Setup:
In pairs, Student A is the Sage; Student B is the Scribe. Students fold a sheet of paper in half and each writes his/her name on one half.

1. The Sage tells the Scribe how to simplify the radical in problem 1.

2. The Scribe records the Sage's instructions step-by-step in writing.

3. If the Sage is correct, the Scribe praises the Sage. Otherwise, the Scribe coaches, then praises.

4. Students switch roles for the next problem.

▶ **Structure**
• Sage-N-Scribe

▶ **Materials**
• Transparency 7.1.4
• 1 sheet of paper and pencil per pair

1 NAME MY SET(S)

Structure: RallyCoach

Classify each number with the appropriate set(s) of numbers to which it belongs.

1. 0

2. $\dfrac{16}{3}$

3. $\dfrac{-8}{4}$

4. $\sqrt{5}$

5. $-\sqrt{10}$

6. $-\sqrt{25}$

7. $\sqrt{49}$

8. 5.3

Answers:
1. whole, integer, rational, real
2. rational, real
3. integer, rational, real
4. irrational, real
5. irrational, real
6. integer, rational, real
7. natural, whole, integer, rational, real
8. rational, real

Cooperative Learning and Algebra 1: Becky Bride
Kagan Publishing • 1 (800) 933-2667 • www.KaganOnline.com

SIMPLIFY MY RADICAL

RADICALS
$\sqrt{20}$

Structure: Sage-N-Scribe

Simplify each radical showing your work. Assume all variables are positive numbers.

1. $\sqrt{20}$

2. $\sqrt{104}$

3. $\sqrt{27}$

4. $\sqrt{54}$

5. $3\sqrt{18}$

6. $-2\sqrt{60}$

7. $3\sqrt{90}$

8. $-2\sqrt{32}$

9. $\sqrt{x^3}$

10. $\sqrt{x^4 y^5}$

11. $\sqrt{4x^2 y^4}$

12. $\sqrt{9x^8 y^7}$

Answers:

1. $2\sqrt{5}$ 2. $2\sqrt{26}$
3. $3\sqrt{3}$ 4. $3\sqrt{6}$
5. $9\sqrt{2}$ 6. $-4\sqrt{15}$
7. $9\sqrt{10}$ 8. $-8\sqrt{2}$
9. $x\sqrt{x}$ 10. $x^2 y^2 \sqrt{y}$
11. $2xy^2$ 12. $3x^4 y^3 \sqrt{y}$

ACTIVITY 3

SIMPLIFY MY RADICAL FRACTION

Structure: Sage-N-Scribe

Simplify each radical expression.

1. $\sqrt{\dfrac{3}{5}}$

2. $\sqrt{\dfrac{5}{6}}$

3. $\sqrt{\dfrac{3}{7}}$

4. $\sqrt{\dfrac{5}{11}}$

Answers:

1. $\dfrac{\sqrt{15}}{5}$

2. $\dfrac{\sqrt{30}}{6}$

3. $\dfrac{\sqrt{21}}{7}$

4. $\dfrac{\sqrt{55}}{11}$

Cooperative Learning and Algebra 1: Becky Bride
Kagan Publishing • 1 (800) 933-2667 • www.KaganOnline.com

ACTIVITY
4
SIMPLIFY MY RADICAL
FRACTION—ADVANCED

Structure: Sage-N-Scribe

Simplify each radical expression.

1. $\dfrac{3\sqrt{6}}{9\sqrt{10}}$

2. $\dfrac{12\sqrt{8}}{4\sqrt{6}}$

3. $\dfrac{12\sqrt{24}}{18\sqrt{32}}$

4. $\dfrac{15\sqrt{27}}{20\sqrt{18}}$

Answers:

1. $\dfrac{\sqrt{15}}{15}$

2. $2\sqrt{3}$

3. $\dfrac{\sqrt{3}}{3}$

4. $\dfrac{3\sqrt{6}}{8}$

LESSON 2
OPERATIONS WITH RADICALS

This lesson has students add, subtract, and multiply radicals. The polynomial chapter is reinforced in this lesson because students will be using the distributive property in Activity 4 and will be multiplying binomials that contain radicals in Activity 5.

ADD/SUBTRACT MY RADICALS

▶ **Structure**
• Sage-N-Scribe

▶ **Materials**
• Transparency 7.2.1
• 1 sheet of paper and pencil per pair

Algebraic

Setup:
In pairs, Student A is the Sage; Student B is the Scribe. Students fold a sheet of paper in half and each writes his/her name on one half.

1. The Sage tells the Scribe how to add or subtract the radicals in problem 1.

2. The Scribe records the Sage's instructions step-by-step in writing.

3. If the Sage is correct, the Scribe praises the Sage. Otherwise, the Scribe coaches, then praises.

4. Students switch roles for the next problem.

ADD/SUBTRACT MY RADICALS—ADVANCED

▶ **Structure**
• Sage-N-Scribe

▶ **Materials**
• Transparency 7.2.2
• 1 sheet of paper and pencil per pair

Algebraic

Setup:
In pairs, Student A is the Sage; Student B is the Scribe. Students fold a sheet of paper in half and each writes his/her name on one half.

1. The Sage tells the Scribe how to add or subtract the radicals in problem 1.

2. The Scribe records the Sage's instructions step-by-step in writing.

3. If the Sage is correct, the Scribe praises the Sage. Otherwise, the Scribe coaches, then praises.

4. Students switch roles for the next problem.

Cooperative Learning and Algebra 1: Becky Bride
Kagan Publishing • 1 (800) 933-2667 • www.KaganOnline.com

3

MULTIPLY MY RADICALS

Algebraic

Setup:
In pairs, Student A is the Sage; Student B is the Scribe. Students fold a sheet of paper in half and each writes his/her name on one half.

1. The Sage tells the Scribe how to multiply the radicals in problem 1.

2. The Scribe records the Sage's instructions step-by-step in writing.

3. If the Sage is correct, the Scribe praises the Sage. Otherwise, the Scribe coaches, then praises.

4. Students switch roles for the next problem.

▶ **Structure**
• Sage-N-Scribe

▶ **Materials**
• Transparency 7.2.3
• 1 sheet of paper and pencil per pair

4

DISTRIBUTE MY RADICAL

Algebraic

Setup:
In pairs, Student A is the Sage; Student B is the Scribe. Students fold a sheet of paper in half and each writes his/her name on one half.

1. The Sage tells the Scribe how to use the distributive property to multiply the radicals in problem 1.

2. The Scribe records the Sage's instructions step-by-step in writing.

3. If the Sage is correct, the Scribe praises the Sage. Otherwise, the Scribe coaches, then praises.

4. Students switch roles for the next problem.

▶ **Structure**
• Sage-N-Scribe

▶ **Materials**
• Transparency 7.2.4
• 1 sheet of paper and pencil per pair

5

MULTIPLY MY RADICALS—ADVANCED

Algebraic

Setup:
In pairs, Student A is the Sage; Student B is the Scribe. Students fold a sheet of paper in half and each writes his/her name on one half.

1. The Sage tells the Scribe how to multiply the radicals in problem 1.

2. The Scribe records the Sage's instructions step-by-step in writing.

3. If the Sage is correct, the Scribe praises the Sage. Otherwise, the Scribe coaches, then praises.

4. Students switch roles for the next problem.

▶ **Structure**
• Sage-N-Scribe

▶ **Materials**
• Transparency 7.2.5
• 1 sheet of paper and pencil per pair

Chapter 7: Radicals
Lesson Two

ADD/SUBTRACT MY RADICALS

Structure: Sage-N-Scribe

Simplify each expression.

1. $3\sqrt{7} + 2\sqrt{5} - 4\sqrt{7}$

2. $4\sqrt{2} + 2\sqrt{2} - 7\sqrt{5}$

3. $-8\sqrt{6} - 2\sqrt{6} - 7\sqrt{3}$

4. $-9\sqrt{11} - 2\sqrt{10} + 7\sqrt{11}$

5. $3\sqrt{x} - 5\sqrt{y} + 9\sqrt{x}$

6. $4\sqrt{m} + 3\sqrt{m} - 8\sqrt{w}$

Answers:
1. $-\sqrt{7} + 2\sqrt{5}$
2. $6\sqrt{2} - 7\sqrt{5}$
3. $-10\sqrt{6} - 7\sqrt{3}$
4. $-2\sqrt{11} - 2\sqrt{10}$
5. $12\sqrt{x} - 5\sqrt{y}$
6. $7\sqrt{m} - 8\sqrt{w}$

Cooperative Learning and Algebra 1: Becky Bride
Kagan Publishing • 1 (800) 933-2667 • www.KaganOnline.com

ACTIVITY 2
ADD/SUBTRACT MY RADICALS—ADVANCED

Structure: Sage-N-Scribe

Simplify each expression. Assume all variables are positive numbers.

1. $3\sqrt{27} - 2\sqrt{18} + 5\sqrt{8}$

2. $4\sqrt{24} - 2\sqrt{54} + 5\sqrt{32}$

3. $4\sqrt{28} + 5\sqrt{40} - 3\sqrt{63}$

4. $7\sqrt{45} - 5\sqrt{20} + \sqrt{80}$

5. $3\sqrt{x^3} - 5\sqrt{y^4} + 6\sqrt{x^3}$

6. $\sqrt{w^5} - 7\sqrt{m^4} + \sqrt{m^4}$

Answers:
1. $9\sqrt{3} + 4\sqrt{2}$
2. $2\sqrt{6} + 20\sqrt{2}$
3. $-\sqrt{7} + 10\sqrt{10}$
4. $15\sqrt{5}$
5. $9x\sqrt{x} - 5y^2$
6. $w^2\sqrt{w} - 6m^2$

ACTIVITY
3

MULTIPLY MY RADICALS

Structure: Sage-N-Scribe

Simplify each radical expression. Assume all variables are positive numbers.

1. $\left(6\sqrt{3}\right)\left(5\sqrt{2}\right)$ 2. $\left(-\sqrt{6}\right)\left(5\sqrt{15}\right)$

3. $\left(-2\sqrt{10}\right)\left(3\sqrt{15}\right)$ 4. $\left(-4\sqrt{12}\right)\left(2\sqrt{8}\right)$

5. $\left(3\sqrt{4x^3}\right)\left(\sqrt{6x^5}\right)$ 6. $\left(\sqrt{8y^5}\right)\left(\sqrt{y^4}\right)$

Answers:
1. $30\sqrt{6}$
2. $-15\sqrt{10}$
3. $-30\sqrt{6}$
4. $-32\sqrt{6}$
5. $6x^4\sqrt{6}$
6. $2y^4\sqrt{2y}$

Cooperative Learning and Algebra 1: Becky Bride
Kagan Publishing • 1 (800) 933-2667 • www.KaganOnline.com

Structure: Sage-N-Scribe

Simplify each radical expression. Assume all variables are positive numbers.

1. $\sqrt{3}\left(3\sqrt{2} - 4\sqrt{8}\right)$ 2. $2\sqrt{6}\left(4\sqrt{2} - \sqrt{3}\right)$

3. $2\sqrt{12}\left(3\sqrt{2} - \sqrt{3}\right)$ 4. $3\sqrt{10}\left(5\sqrt{2} - 4\sqrt{5}\right)$

5. $\left(3\sqrt{x}\right)\left(2\sqrt{x} - \sqrt{5}\right)$ 6. $\left(2\sqrt{y}\right)\left(3\sqrt{y^2} - 4\sqrt{y}\right)$

Answers:
1. $-5\sqrt{6}$
2. $16\sqrt{3} - 6\sqrt{2}$
3. $12\sqrt{6} - 12$
4. $30\sqrt{5} - 60\sqrt{2}$
5. $6x - 3\sqrt{5x}$
6. $6y\sqrt{y} - 8y$

Cooperative Learning and Algebra 1: Becky Bride
Kagan Publishing • 1 (800) 933-2667 • www.KaganOnline.com

MULTIPLY MY RADICALS—ADVANCED

Structure: Sage-N-Scribe

Simplify each radical expression. Assume all variables are positive numbers.

1. $\left(2\sqrt{3}+1\right)\left(\sqrt{3}-1\right)$

2. $\left(4\sqrt{5}+3\right)\left(2\sqrt{5}-1\right)$

3. $\left(7\sqrt{6}-3\right)\left(3\sqrt{6}-2\right)$

4. $\left(\sqrt{7}+3\right)\left(5\sqrt{7}+4\right)$

5. $\left(6\sqrt{x}+1\right)\left(2\sqrt{x}-3\right)$

6. $\left(5-2\sqrt{y}\right)\left(3+\sqrt{y}\right)$

Answers:
1. $5-\sqrt{3}$
2. $37+2\sqrt{5}$
3. $132-23\sqrt{6}$
4. $47+19\sqrt{7}$
5. $12x-16\sqrt{x}-3$
6. $15-\sqrt{y}-2y$

Cooperative Learning and Algebra 1: Becky Bride
Kagan Publishing • 1 (800) 933-2667 • www.KaganOnline.com

LESSON 3
APPLICATIONS

In this lesson students will be finding the area of geometric figures and will need to know or have the formulas for these figures. The Pythagorean Theorem is also introduced with applications in a subsequent activity.

1 AREA AND RADICALS

Algebraic with Geometric Connection

Setup:
In pairs, Student A is the Sage; Student B is the Scribe. Students fold a sheet of paper in half and each writes his/her name on one half.

1. The Sage tells the Scribe how to solve problem 1.

2. The Scribe records the Sage's instructions step-by-step in writing.

3. If the Sage is correct, the Scribe praises the Sage. Otherwise, the Scribe coaches, then praises.

4. Students switch roles for the next problem.

▶ **Structure**
• Sage-N-Scribe

▶ **Materials**
• 1 Blackline 7.3.1 per pair
• 1 sheet of paper and pencil per pair

2 PYTHAGOREAN THEOREM

Algebraic with Geometric Connection

Setup:
In pairs, Student A is the Sage; Student B is the Scribe. Students fold a sheet of paper in half and each writes his/her name on one half.

1. The Sage tells the Scribe how to solve problem 1.

2. The Scribe records the Sage's instructions step-by-step in writing.

3. If the Sage is correct, the Scribe praises the Sage. Otherwise, the Scribe coaches, then praises.

4. Students switch roles for the next problem.

▶ **Structure**
• Sage-N-Scribe

▶ **Materials**
• 1 Blackline 7.3.2 per pair
• 1 sheet of paper and pencil per pair

ACTIVITY

3 APPLICATIONS WITH THE PYTHAGOREAN THEOREM

▶ **Structure**
• Sage-N-Scribe

▶ **Materials**
• 1 Blackline 7.3.3 per pair
• 1 sheet of paper and pencil per pair

Algebraic with Geometric Connection

Setup:
In pairs, Student A is the Sage; Student B is the Scribe. Students fold a sheet of paper in half and each writes his/her name on one half.

1. The Sage tells the Scribe how to solve problem 1.

2. The Scribe records the Sage's instructions step-by-step in writing.

3. If the Sage is correct, the Scribe praises the Sage. Otherwise, the Scribe coaches, then praises.

4. Students switch roles for the next problem.

ACTIVITY

4 WHAT DID WE LEARN?

▶ **Structure**
• RoundTable Consensus

▶ **Materials**
• 1 large sheet of paper per team
• different color pen or pencil for each student on the team

Synthesis via Graphic Organizer

1. Each teammate signs his/her name in the upper right corner of the team paper with the color pen/pencil he/she is using.

2. One teammate writes "Radicals" in the center of the team paper in a rectangle.

3. Teammate 1 shares with the team one core concept he/she learned in the unit.

4. The student checks for consensus.

5. The teammates show agreement or lack of agreement with thumbs up or down.

6. If there is agreement, the students celebrate and the teammate records the core concept on the graphic organizer connecting it with a line to the main idea, "Radicals." If not, teammates discuss the response until there is agreement and then they celebrate.

7. Play continues with the next student's core concept until all core concepts are exhausted.

8. Repeat steps 3-7, with teammates adding details to each core concept and making bridges between related ideas.

Cooperative Learning and Algebra 1: Becky Bride
Kagan Publishing • 1 (800) 933-2667 • www.KaganOnline.com

ACTIVITY
1 **AREA AND RADICALS**

Structure: Sage-N-Scribe

Find the area of each figure below.

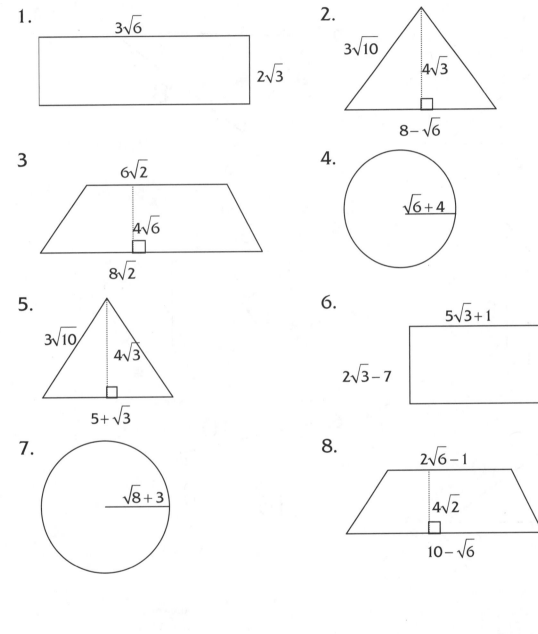

1.
$3\sqrt{6}$
$2\sqrt{3}$

2.
$3\sqrt{10}$
$4\sqrt{3}$
$8-\sqrt{6}$

3
$6\sqrt{2}$
$4\sqrt{6}$
$8\sqrt{2}$

4.
$\sqrt{6}+4$

5.
$3\sqrt{10}$
$4\sqrt{3}$
$5+\sqrt{3}$

6.
$5\sqrt{3}+1$
$2\sqrt{3}-7$

7.
$\sqrt{8}+3$

8.
$2\sqrt{6}-1$
$4\sqrt{2}$
$10-\sqrt{6}$

Answers:

1. $18\sqrt{2}$ 2. $16\sqrt{3}-6\sqrt{2}$
3. $56\sqrt{3}$ 4. $\left(22+8\sqrt{6}\right)\pi$
5. $10\sqrt{3}+6$ 6. $23-33\sqrt{3}$
7. $\left(17+12\sqrt{2}\right)\pi$ 8. $4\sqrt{3}+18\sqrt{2}$

ACTIVITY

2 PYTHAGOREAN THEOREM

Structure: Sage-N-Scribe

Find the missing side. All answers need to be in simplest radical form. Diagrams are not drawn to scale.

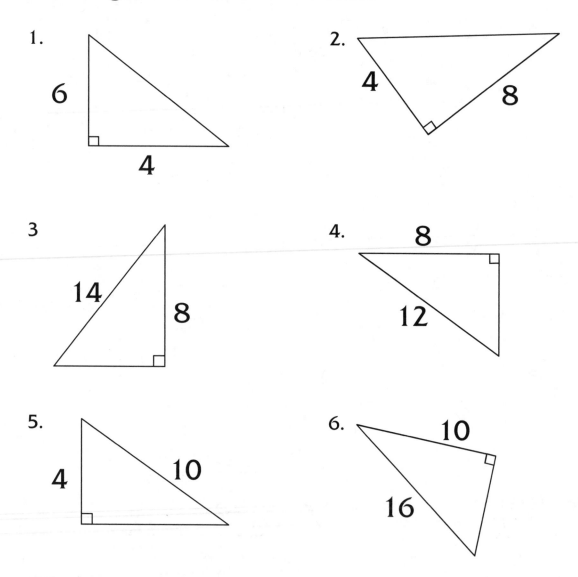

1.

6

4

2.

4

8

3

14

8

4.

8

12

5.

4

10

6.

10

16

Answers:
1. $2\sqrt{13}$
2. $4\sqrt{5}$
3. $2\sqrt{33}$
4. $4\sqrt{5}$
5. $2\sqrt{21}$
6. $2\sqrt{39}$

Cooperative Learning and Algebra 1: Becky Bride
Kagan Publishing • 1 (800) 933-2667 • www.KaganOnline.com

APPLICATIONS WITH THE PYTHAGOREAN THEOREM

ACTIVITY 3

√20 RADICALS

Structure: Sage-N-Scribe

Round each answer to the nearest tenth.

1. Jose's school is two miles north of his house, and the soccer fields are three miles east of his house. What is the distance from Jose's school to the soccer fields?

2. A 10-foot ladder is placed against a house. If the base of the ladder is three and a half feet from the base of the house, how high does the ladder reach on the house?

3. One end of a guy wire will attach to a telephone 10 feet above the ground. The other end of the wire will attach to a bracket 30 feet from the base of the pole. How long of a wire is needed?

4. The location of Joan's house, school, and dance studio form a right triangle with the house, school, and dance studio at the vertices of the triangle. Joan's house is located at the vertex of the right triangle where the right angle is. If the distance from the dance studio to the school is twelve miles and the distance from her house to the studio is six miles, find the distance from her house to the school.

Answers:
1. 3.6 miles
2. 9.4 feet
3. 31.6 feet
4. 10.4 miles

CHAPTER 8

QUADRATIC FUNCTIONS

The focus of this chapter is quadratic functions. The first lesson develops the vocabulary necessary to do the exploratory activities in Lesson 2 which explores the role of "a," "h," and "k" in the vertex form of a quadratic function. Students use this knowledge to graph quadratic functions given in vertex form. The students will change a quadratic function from vertex form into standard form. By doing this, students see that the two forms describe the same curve. An exploratory activity follows so students will discover how to find the x-value of the vertex of a parabola whose equation is in standard form. Activities follow, requiring students to find the vertex of a parabola and graph a parabola whose equation is in standard form. Lesson 3 establishes the connection between the quadratic formula and x-intercepts and has students work with the quadratic formula and the discriminant. The application activity has students find maximum and minimum values and x-intercepts in the context of real-world problems.

LESSON 1 VOCABULARY DEVELOPMENT

ACTIVITY 1: What Do I Look Like?
ACTIVITY 2: Can You Define Me?
ACTIVITY 3: Am I a Quadratic Function?
ACTIVITY 4: Use Your New Vocabulary
ACTIVITY 5: Use Your New Vocabulary: Part 2
ACTIVITY 6: What Did We Learn?

LESSON 2 GRAPHING QUADRATIC FUNCTIONS

ACTIVITY 1: Exploring the Role of "a"
ACTIVITY 2: What Did I Do?
ACTIVITY 3: Exploring the Role of "h"
ACTIVITY 4: What Did I Do?
ACTIVITY 5: Exploring the Role of "k"
ACTIVITY 6: What Did I Do?
ACTIVITY 7: Graph Me (Vertex Form)
ACTIVITY 8: Write My Equation
ACTIVITY 9: Rewrite Me in Standard Form
ACTIVITY 10: What are the Values of a, b, and c?
ACTIVITY 11: Exploring the Vertex (Standard Form)
ACTIVITY 12: Find My Vertex
ACTIVITY 13: Graph Me (Standard Form)
ACTIVITY 14: Rewrite Me in Vertex Form

LESSON 3 SOLVING QUADRATIC FUNCTIONS

ACTIVITY 1: Exploring the Connection: Quadratic Formula and X-Intercepts
ACTIVITY 2: Find My Solution(s)
ACTIVITY 3: Find My Solution(s) and Simplify
ACTIVITY 4: Exploring the Discriminant
ACTIVITY 5: How Many Solutions Do I Have?
ACTIVITY 6: Applications
ACTIVITY 7: What Did We Learn?

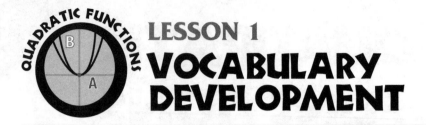

LESSON 1
VOCABULARY DEVELOPMENT

This lesson develops the vocabulary that will be used in the exploratory activities in Lesson 2. This lesson lays the vocabulary foundation for functions other than quadratic because these terms are used for other functions. That is why the graphs are not all quadratic but involve other functions as well. A vertical stretch pulls the graph closer to the y-axis. The students will want to describe this type of transformation as making the graph "skinnier," but that description is not accurate for the square root function. Processing each definition as the class finishes defining each word is the best use of the processing activities. Immediate processing has a greater chance of solidifying the definition in the student's mind.

WHAT DO I LOOK LIKE?

▶ **Structure**
• Solo-Pair Consensus-Team Consensus

▶ **Materials**
• 1 Blackline 8.1.1 per student
• 1 sheet of graph paper and pen/pencil per student

Exploratory

Solo

1. Have each student complete the investigation individually.

Pair Consensus

2. For each problem on the investigation, each student shares with his/her partner, using RallyRobin, his/her response. They discuss the problems that they disagree on, trying to come to consensus on the correct response. For the problems they can't reach consensus on, they mark these so they can focus on them during the team phase. Encourage the students to add to their responses if their partner verbalized an understanding they did not see.

Team Consensus

3. Each pair shares their responses, using RallyRobin, with the other pair in their team, augmenting their responses if necessary. When the teams are through sharing, each student should have a detailed, complete summary.

Cooperative Learning and Algebra 1: Becky Bride
Kagan Publishing • 1 (800) 933-2667 • www.KaganOnline.com

2

CAN YOU DEFINE ME?

Exploratory

Solo

1. Individually, each student defines each vocabulary word by comparing and contrasting the examples and counterexamples. (Students could do this part for homework.)

Team with RoundRobin Sharing and RoundTable Recording

2. In turn, each student reads to his/her team the definition he/she wrote.

3. The team discusses how to define the vocabulary word, based on the definitions just shared. The team must come to consensus on how to define the term.

4. Once the team reaches consensus, student 1 records the team definition on the team paper.

5. Repeat steps 2-4 for each vocabulary word, rotating the recording.

Class

6. Choose a team and a student at that team to read his/her team's definition of the first vocabulary word. (Team and student spinners work well here.)

7. Write the definition on the board. Ask the class if they agree with part 1, then part 2, and finally part 3 of the definition, reworking what they want to change.

8. Once everyone has agreed, including the teacher, then it is recorded as the definition

▶ **Structure**
• Solo Team Class

▶ **Materials**
• 1 Blackline 8.1.2 and pencil per student
• 1 sheet of paper and pencil per team

the class will use. You may find that lower-level classes prefer a somewhat longer definition if it has more meaning for them. Honors classes will try to be as concise as possible.

9. Repeat steps 6-8 for the remainder of the words.

Tip: Processing each definition after it is defined with the activities that follow helps to solidify the definition in the minds of the students. Activity 3 is for quadratic function. Activity 4 is for definitions 2-7. Activity 5 is for definitions 8-11.

3

AM I A QUADRATIC FUNCTION?

Algebraic

1. Teacher poses a multiple problems with Transparency 8.1.3.

2. In pairs, students take turns stating whether the function is quadratic, orally.

▶ **Structure**
• RallyRobin

▶ **Materials**
• Transparency 8.1.3

ACTIVITY

4 USE YOUR NEW VOCABULARY

▶ **Structure**
· RallyRobin

▶ **Materials**
· Transparency 8.1.4

Exploratory

1. Teacher poses multiple problems with Transparency 8.1.4.

2. In pairs, students take turns practicing a new definition, orally.
 Round 1: Vertex
 Round 2: Axis of Symmetry
 Round 3: Maximum point and maximum value
 Round 4: Minimum point and minimum value

Answers:
Round 1:
 1. (4, −7)
 2. (3, 10)
 3. (1, 7)
 4. (−1, −4)

Round 2:
 1. x = 4
 2. x = 3
 3. x = 1
 4. x = − 1

Round 3:
 1. none
 2. (3, 10); 10
 3. (1, 7); 7
 4. none

Round 4:
 1. (4, −7); −7
 2. none
 3. none
 4. (−1, −4); −4

ACTIVITY

5 USE YOUR NEW VOCABULARY: PART 2

▶ **Structure**
· RallyRobin

▶ **Materials**
· Transparency 8.1.5

Exploratory

1. Teacher poses multiple problems with Transparency 8.1.5.

2. In pairs, students take turns practicing a new definition, orally.
 Round 1: Vertical stretch and vertical compression
 Round 2: Horizontal shift
 Round 3: Vertical shift

Answers:
Round 1:
 1. vertical stretch
 2. vertical compression
 3. vertical compression
 4. vertical stretch

Round 2:
 1. horizontal shift
 2. none
 3 . horizontal shift
 4. horizontal shift

Round 3:
 1. none
 2. vertical shift
 3. vertical shift
 4. vertical shift

Cooperative Learning and Algebra 1: Becky Bride
Kagan Publishing • 1 (800) 933-2667 • www.KaganOnline.com

ACTIVITY

6   WHAT DID WE LEARN?

Synthesis via Graphic Organizer

1. Each teammate signs his/her name in the upper-right corner of the team paper with the color pen/pencil he/she is using.

2. One teammate writes "Quadratic Functions" in the center of the team paper in a rectangle.

3. Teammate 1 shares with the team one core concept he/she learned in the entire unit.

4. The student checks for consensus.

5. The teammates show agreement or lack of agreement with thumbs up or down.

6. If there is agreement, the students celebrate and the teammate records the core concept on the graphic organizer connecting it with a line to the main idea, "Quadratic Functions." If not, teammates discuss the response until there is agreement and then they celebrate.

7. Play continues with the next student's core concept until all core concepts are exhausted.

8. Repeat steps 3-7, with teammates adding details to each core concept and making bridges between related ideas.

▶ **Structure**
· RoundTable Consensus

▶ **Materials**
· 1 large sheet of paper per team
· different color pen or pencil for each student on the team

ACTIVITY

1 WHAT DO I LOOK LIKE?

1. Complete the table below.

2. Graph the function $y = x^2$ on the graph below using the table to the left.

x	$y = x^2$	(x, y)
–3		
–2		
–1		
0		
1		
2		
3		

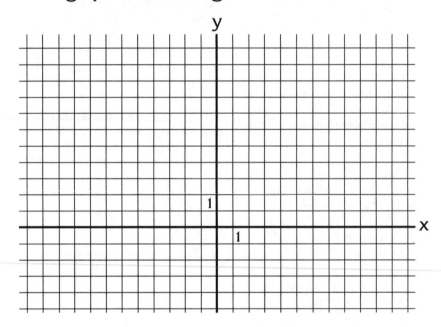

3. Describe the graph.

4. Describe the symmetry of the graph.

5. Why aren't any y–values negative?

This graph is called a "parabola." All quadratic functions have a shape similar to this. The focus of the next unit will be on quadratic functions and their graphs.

Cooperative Learning and Algebra 1: Becky Bride
Kagan Publishing • 1 (800) 933-2667 • www.KaganOnline.com

CAN YOU DEFINE ME?

Compare the examples to the counterexamples and write a definition of the term or phrase based on your comparisons.

1. Quadratic Function

Examples	**Counterexamples**

$f(x) = x^2$ $m(x) = 3x + 3$

$g(x) = x^2 + 3x - 1$ $n(x) = 5$

$h(x) = 5x + x^2$ $p(x) = x^3 + 2 + x^2$

$k(x) = 4 - 3x^2$ $w(x) = 3x^4 - 5$

$l(x) = \dfrac{2}{3}x^2 + 5$ $z(x) = \dfrac{5}{x} + 2$

2. Vertex

Examples

Point A is an example of a vertex.

Counterexamples

Point A is not a vertex.

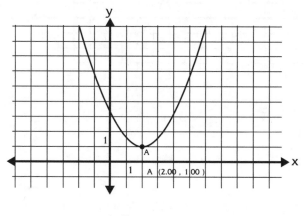

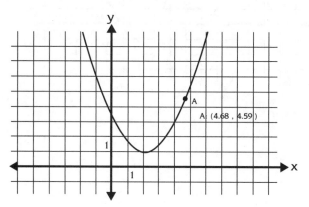

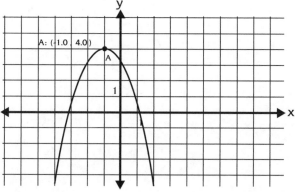

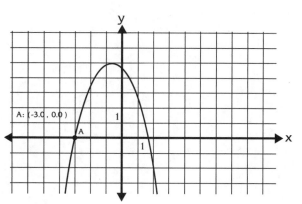

Cooperative Learning and Algebra 1: Becky Bride
Kagan Publishing • 1 (800) 933-2667 • www.KaganOnline.com

ACTIVITY

2 CAN YOU DEFINE ME?

3. Axis of symmetry

Examples

Counterexamples

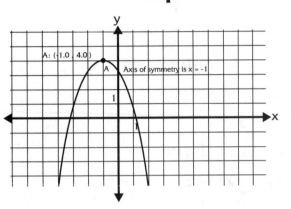

A: (-1.0 , 4.0)

Axis of symmetry is x = -1

1

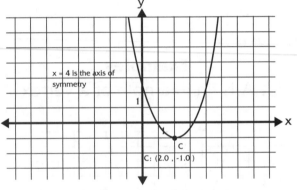

A: (-1.0 , 4.0)

y = 2 is not an axis of symmetry

1

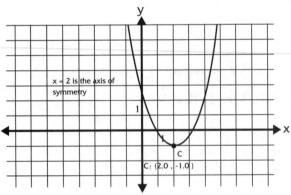

x = 2 is the axis of symmetry

1

C: (2.0 , -1.0)

x = 4 is the axis of symmetry

1

C: (2.0 , -1.0)

Cooperative Learning and Algebra 1: Becky Bride
Kagan Publishing • 1 (800) 933-2667 • www.KaganOnline.com

ACTIVITY

2 CAN YOU DEFINE ME?

4. Maximum point

Examples	**Counterexamples**
Point A is the maximum point.	Point B is not the maximum point.

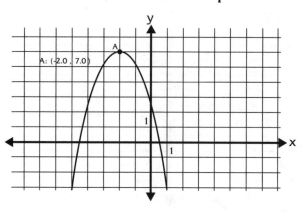
A: (-2.0 , 7.0)

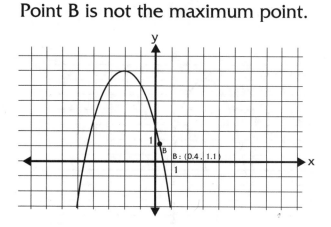
B: (0.4 , 1.1)

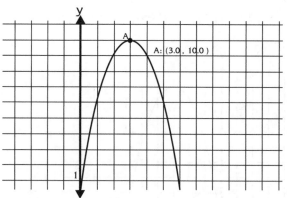
A: (3.0 , 10.0)

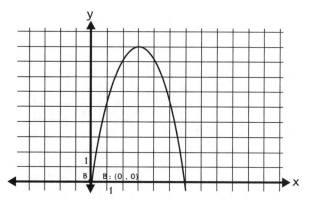
B: (0 , 0)

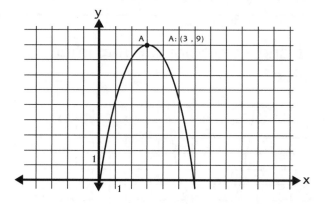
A: (3 , 9)

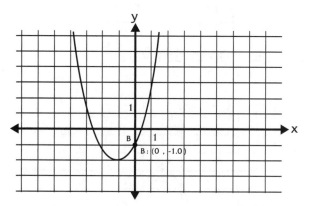
B: (0 , -1.0)

ACTIVITY

2 **CAN YOU DEFINE ME?**

5.   Maximum value

Examples	**Counterexamples**
7 is the maximum value	0.4 and 1.1 are not the maximum value

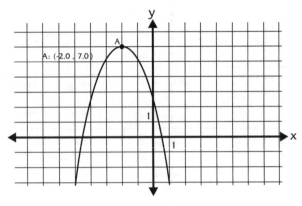

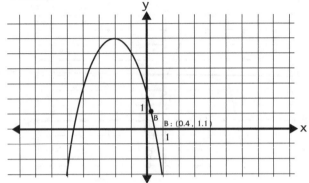

10 is the maximum value	1 and 0 are not the maximum value

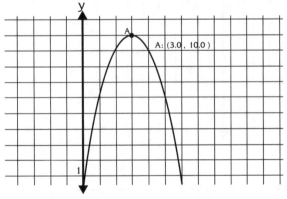

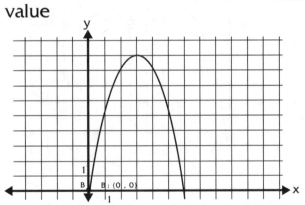

0 is the maximum value	−1 and -2 are not the maximum value

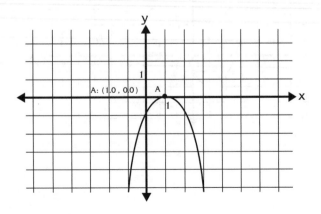

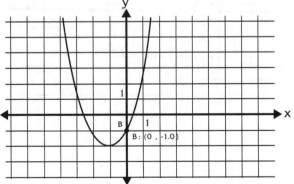

Cooperative Learning and Algebra 1: Becky Bride
Kagan Publishing • 1 (800) 933-2667 • www.KaganOnline.com

ACTIVITY

2 CAN YOU DEFINE ME?

6. Minimum Point

Examples	**Counterexamples**
Point E is a minimum point.	Point F is not a minimum point.

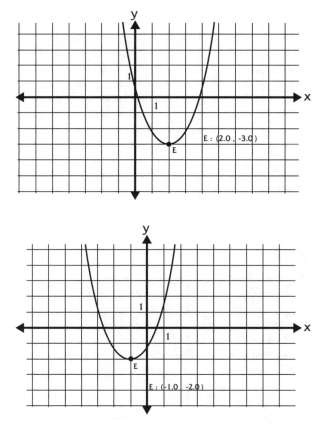

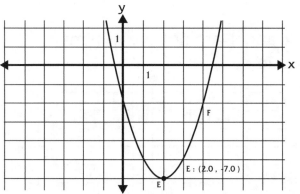

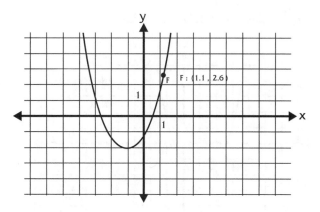

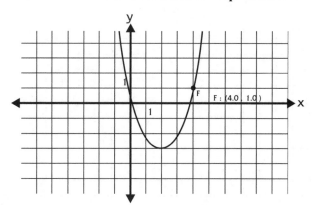

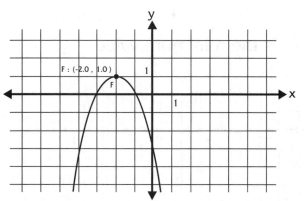

ACTIVITY

2

CAN YOU DEFINE ME?

7. Minimum value

Examples	**Counterexamples**

−3 is the minimum value

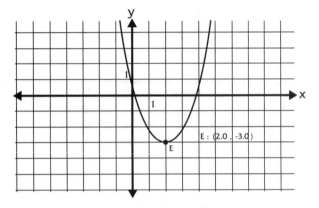

1 and 4 are not the minimum value

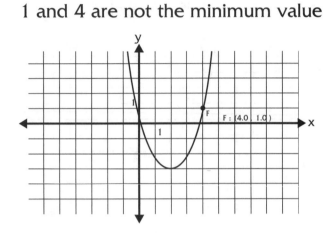

−2 is the minimum value

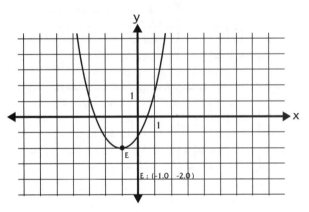

1.1 and 2.6 are not the minimum value

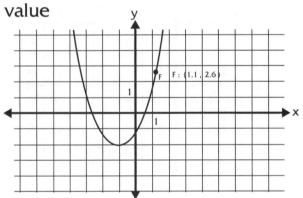

−7 is the minimum value

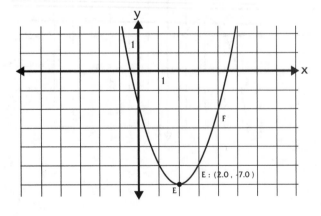

−2 and 1 are not the minimum value

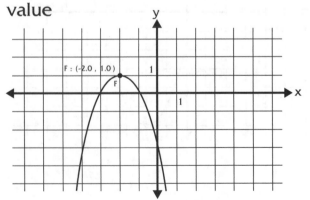

Cooperative Learning and Algebra 1: Becky Bride
Kagan Publishing • 1 (800) 933-2667 • www.KaganOnline.com

ACTIVITY
2

CAN YOU DEFINE ME?

8. Vertical Stretch

Examples	**Counterexamples**
In each example, graph B is a vertical stretch compared to graph A.	In each example, graph B is not a vertical of stretch compared to graph A.

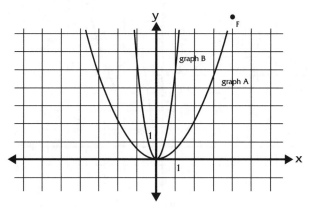

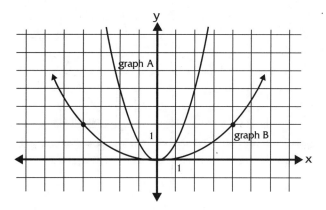

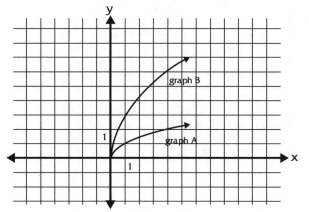

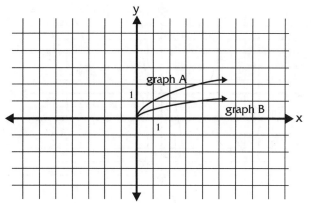

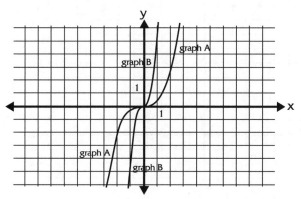

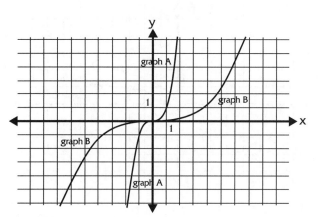

ACTIVITY

2 CAN YOU DEFINE ME?

QUADRATIC FUNCTIONS

9. Vertical Compression

Examples	**Counterexamples**
In each graph, graph B is a vertical compression compared to graph A.	In each graph, graph B is not a vertical compression compared to graph A.

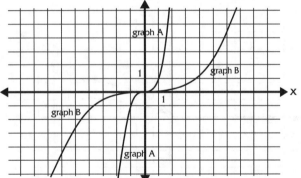

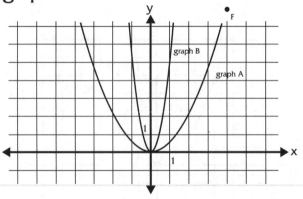

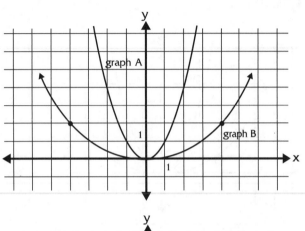

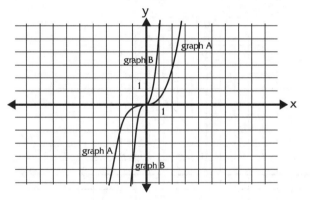

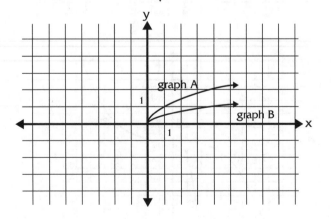

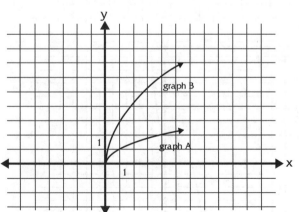

Cooperative Learning and Algebra 1: Becky Bride
Kagan Publishing • 1 (800) 933-2667 • www.KaganOnline.com

ACTIVITY

2 CAN YOU DEFINE ME?

10. Horizontal Shift

Examples	**Counterexamples**
In each example, graph B is a horizontal shift of graph A.	In each example, graph B is not a horizontal shift of graph A.

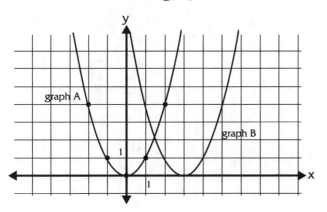

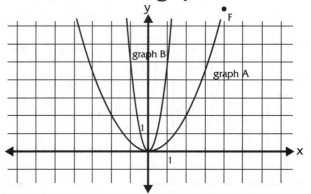

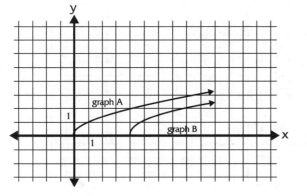

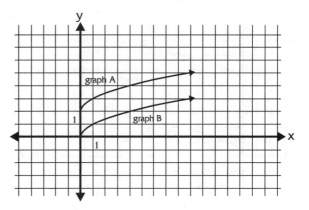

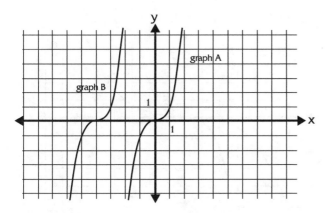

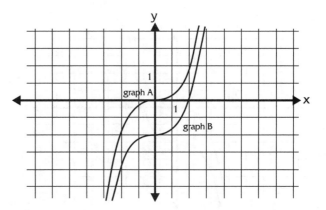

ACTIVITY

2 CAN YOU DEFINE ME?

11. Vertical Shift

Examples	**Counterexamples**
In each example, graph B is a vertical shift of Graph A.	In each example, graph B is not a vertical shift of graph A.

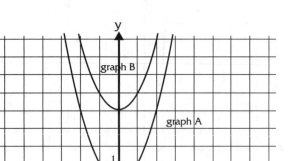

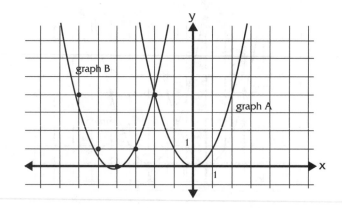

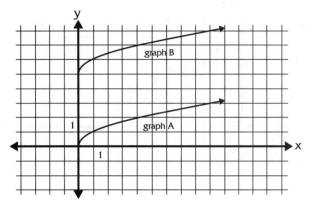

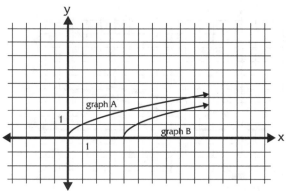

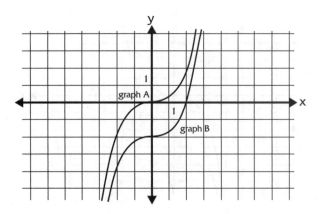

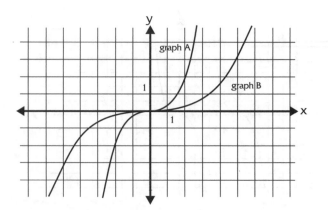

Cooperative Learning and Algebra 1: Becky Bride
Kagan Publishing • 1 (800) 933-2667 • www.KaganOnline.com

AM I A QUADRATIC FUNCTION?

Structure: RallyRobin

Determine if each equation is a quadratic function.

1. $4x + 1 = y$

2. $3x^2 - 7x = y$

3. $y - 1 = x - 2x^2$

4. $5x^2 - x^3 + 3 = y$

5. $\dfrac{3}{4}x^2 + 5x - 2 = y$

6. $\dfrac{4}{x^2} + 3x - 9 = y$

Answers:
1. no
2. yes
3. yes
4. no
5. yes
6. no

ACTIVITY
4 **USE YOUR NEW VOCABULARY**

Structure: RallyRobin

Round 1: State the vertex of each parabola.
Round 2: State the axis of symmetry of each parabola.
Round 3: State the maximum point and the maximum value.
Round 4: State the minimum point and the minimum value.
Note: The scale on each axis is 1.

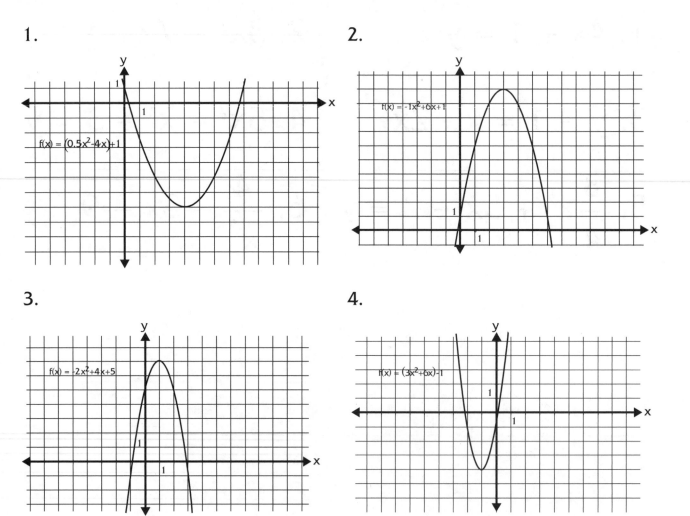

1.

$f(x) = (0.5x^2 - 4x) + 1$

2.

$f(x) = -1x^2 + 6x + 1$

3.

$f(x) = -2x^2 + 4x + 5$

4.

$f(x) = (3x^2 + 6x) - 1$

Answers are on teachers notes (page 390).

Cooperative Learning and Algebra 1: Becky Bride
Kagan Publishing • 1 (800) 933-2667 • www.KaganOnline.com

ACTIVITY 5 USE YOUR NEW VOCABULARY: PART 2

Structure: RallyRobin

Round 1: State whether graph B has a vertical stretch or compression when compared to graph A.

Round 2: State whether graph B has been shifted horizontally when compared to graph A.

Round 3: State whether graph B has been shifted vertically when compared to graph A.

1.

2.

3.

4.

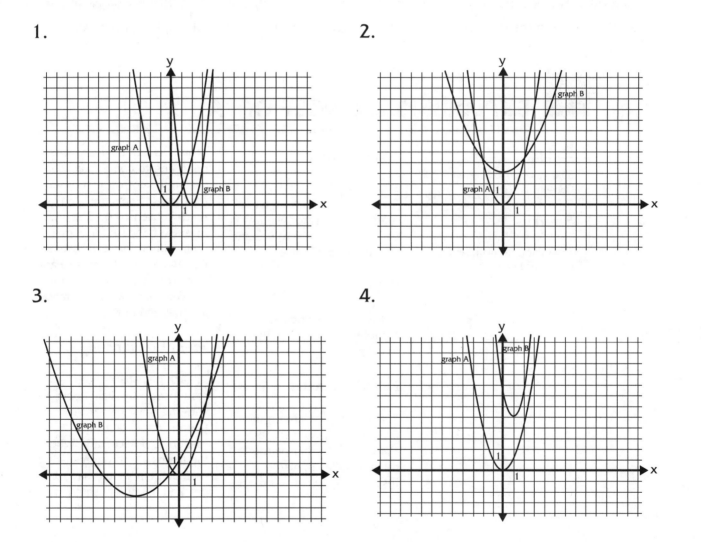

Answers are on teachers notes (page 390).

LESSON 2
GRAPHING QUADRATIC FUNCTIONS

This lesson begins with exploring the vertex form of the equation of a parabola: $y = a(x - h)^2 + k$. Each constant is explored separately, with processing after each exploration. Activity 7 is the culmination of these explorations as students graph quadratic functions that are written in vertex form. Activity 8 reverses this process with students writing the equation of quadratic functions given the graph. Activity 9 has students transform the vertex form of a quadratic equation into standard form, connecting the polynomial unit. Activity 10 has students explore the standard form of a quadratic equation and find the x-value of the vertex by using $\dfrac{-b}{2a}$. This lesson closes with students finding the vertex of a quadratic function in standard form and graphing the function.

ACTIVITY

1 EXPLORING THE ROLE OF "A"

▶ **Structure**
· Solo-Pair Consensus-Team Consensus

▶ **Materials**
· 1 Blackline 8.2.1 per student
· 1 graphing calculator and pen/pencil per student

Exploratory

Setup:
Have students make their graphing calculator window be –10 to 10 for x-values and –10 to 10 for y-values.

Solo

1. Have each student complete the investigation individually.

Pair Consensus

2. For each problem on the investigation, each student shares with his/her partner, using RallyRobin, his/her response. They discuss the problems that they disagree on, trying to come to consensus on the correct response. For the problems they can't reach consensus on, they mark these so they can focus on them during the team phase. Encourage the students to add to their responses if their partner verbalized an understanding they did not see.

Team Consensus

3. Each pair shares their responses using RallyRobin with the other pair in their team, augmenting their responses if necessary. When the teams are through sharing, each student should have a detailed, complete summary.

2 WHAT DID I DO?

Algebraic with Geometric Connection

1. Teacher uses Transparency 8.2.2 to pose multiple questions.

2. In pairs, students take turns stating a) how the graph opens, b) whether the graph was reflected over the x-axis, and c) whether the graph has a vertical stretch or compression, orally.

Answers:
1. a. up; b. no reflection; c. stretch

2. a. down; b. reflection; c. compression

3. a. down; b. reflection; c. compression

4. a. up; b. no reflection; c. stretch

5. a. down; b. reflection; c. compression

6. a. up; b. no reflection; c. stretch

▶ **Structure**
• RallyRobin

▶ **Materials**
• Transparency 8.2.2

3 EXPLORING THE ROLE OF "H"

Exploratory

Setup:
Have students make their graphing calculator window be −10 to 10 for x-values and −10 to 10 for y-values.

Solo

1. Have each student complete the investigation individually.

Pair Consensus

2. For each problem on the investigation, each student shares with his/her partner, using RallyRobin, his/her response. They discuss the problems that they disagree on, trying to come to consensus on the correct response. For the problems they can't reach consensus on, they mark these so they can focus on them during the team phase. Encourage the students to add to their responses if their partner verbalized an understanding they did not see.

Team Consensus

3. Each pair compares their responses to the responses of the other pair in their team, augmenting their responses if necessary. When the teams are through sharing, each student should have a detailed, complete summary.

▶ **Structure**
• Solo-Pair Consensus-Team Consensus

▶ **Materials**
• 1 Blackline 8.2.3 per student
• 1 graphing calculator and pen/pencil per student

ACTIVITY

4 WHAT DID I DO?

▶ **Structure**
· RallyRobin

▶ **Materials**
· Transparency 8.2.2

Algebraic with Geometric Connection

1. Teacher uses Transparency 8.2.2 to pose multiple questions.

2. In pairs, students take turns describing the horizontal shift, orally.

Answers:

1. 4 units right

2. 2 units left

3. none

4. 1 unit right

5. 3 units left

6. none

ACTIVITY

5 EXPLORING THE ROLE OF "K"

▶ **Structure**
· Solo-Pair Consensus-Team Consensus

▶ **Materials**
· 1 Blackline 8.2.4 per student
· 1 graphing calculator and pen/pencil per student

Exploratory

Setup:
Have students make their graphing calculator window be –10 to 10 for x-values and –10 to 10 for y-values.

Solo

1. Have each student complete the investigation individually.

Pair Consensus

2. For each problem on the investigation, each student shares with his/her partner, using RallyRobin, his/her response. They discuss the problems that they disagree on, trying to come to consensus on the correct response. For the problems they can't reach consensus on, they mark these so they can focus on them during the team phase. Encourage the students to add to their responses if their partner verbalized an understanding they did not see.

Team Consensus

3. Each pair shares their responses using RallyRobin with the other pair in their team, augmenting their responses if necessary. When the teams are through sharing, each student should have a detailed, complete summary.

Cooperative Learning and Algebra 1: Becky Bride
Kagan Publishing · 1 (800) 933-2667 · www.KaganOnline.com

6 WHAT DID I DO?

Algebraic with Geometric Connection

1. Teacher uses Transparency 8.2.2 to pose multiple questions.

2. In pairs, students take turns describing the vertical shift, orally.

Answers:

1. 5 units down

2. 4 units up

3. 1 unit up

4. none

5. 6 units up

6. 8 units up

▶ **Structure**
· RallyRobin

▶ **Materials**
· Transparency 8.2.2

7 GRAPH ME (VERTEX FORM)

Graphic

Setup:
In pairs, Student A is the Sage; Student B is the Scribe. Students fold a sheet of paper in half and each writes his/her name on one half.

1. The Sage tells the Scribe how to graph problem 1.

2. The Scribe graphs the parabola using the Sage's instructions.

3. If the Sage is correct, the Scribe praises the Sage. Otherwise, the Scribe coaches, then praises.

4. Students switch roles for the next problem.

▶ **Structure**
· Sage-N-Scribe

▶ **Materials**
· Transparency 8.2.5 and Answer Transparency 8.2.5a
· 1 sheet of graph paper and pencil per pair

Chapter 8: Quadratic Functions

Lesson Two

ACTIVITY

8 WRITE MY EQUATION

Using $y = a(x - h)^2 + k$, students will substitute the vertex coordinates for h and k and the coordinates of the additional point given into x and y. This will leave an equation with "a" as the only unknown. Students will then solve for "a" then rewrite the equation with the values of h, k, and a. This activity may only be appropriate for honors level classes.

▶ **Structure**
• Sage-N-Scribe

▶ **Materials**
• 1 Blackline 8.2.6 per pair
• 1 sheet of paper and pencil per pair

Algebraic

Setup:
In pairs, Student A is the Sage; Student B is the Scribe. Students fold a sheet of paper in half and each writes his/her name on one half.

1. The Sage tells the Scribe how to write the equation for the quadratic function in vertex form for problem 1.

2. The Scribe records the Sage's instructions step-by-step in writing.

3. If the Sage is correct, the Scribe praises the Sage. Otherwise, the Scribe coaches, then praises.

4. Students switch roles for the next problem.

Answers:

1. $y = 3(x - 1)^2 + 2$

2. $y = \frac{1}{2}(x + 3)^2 - 1$

3. $y = \frac{2}{3}(x - 3)^2 + 4$

4. $y = -2(x + 4)^2 + 3$

ACTIVITY

9 REWRITE ME IN STANDARD FORM

▶ **Structure**
• Sage-N-Scribe

▶ **Materials**
• Transparency 8.2.5
• 1 sheet of paper and pencil per pair

Algebraic

Setup:
In pairs, Student A is the Sage; Student B is the Scribe. Students fold a sheet of paper in half and each writes his/her name on one half.

1. The Sage tells the Scribe how to rewrite the quadratic function in problem 1 in standard form.

2. The Scribe records the Sage's instructions step-by-step in writing.

3. If the Sage is correct, the Scribe praises the Sage. Otherwise, the Scribe coaches, then praises.

4. Students switch roles for the next problem.

10 WHAT ARE THE VALUES OF A, B, AND C

Algebraic

1. Teacher poses multiple problems with Transparency 8.2.8.

2. In pairs, students take turns stating the value of a, b, and c in the quadratic function, orally.

▶ **Structure**
• RallyRobin

▶ **Materials**
• Transparency 8.2.8

11 EXPLORING THE VERTEX (GIVEN STANDARD FORM)

Exploratory

Setup:
Have students make their graphing calculator window be –10 to 10 for x-values and –10 to 10 for y-values.

Solo

1. Have each student complete the investigation individually.

Pair Consensus

2. For each problem on the investigation, each student shares with his/her partner, using RallyRobin, his/her response. They discuss the problems that they disagree on, trying to come to consensus on the correct response. For the problems they can't reach consensus on, they mark these so they can focus on them during the team phase. Encourage the students to add to their responses if their partner verbalized an understanding they did not see.

Team Consensus

3. Each pair shares their responses, using RallyRobin, with the other pair in their team, augmenting their responses if necessary. When the teams are through sharing, each student should have a detailed, complete summary.

▶ **Structure**
• Solo-Pair Consensus-Team Consensus

▶ **Materials**
• 1 Blackline 5.2.7 per student
• 1 sheet of graph paper and pen/pencil per student

Chapter 8: Quadratic Functions

Lesson Two

ACTIVITY

12 FIND MY VERTEX

▶ **Structure**
· Sage-N-Scribe

▶ **Materials**
· Transparency 8.2.8
· 1 sheet of paper and pencil per pair

Algebraic

Setup:
In pairs, Student A is the Sage; Student B is the Scribe. Students fold a sheet of paper in half and each writes his/her name on one half.

1. The Sage tells the Scribe how to find the vertex for the parabola in problem 1.

2. The Scribe records the Sage's instructions step-by-step in writing.

3. If the Sage is correct, the Scribe praises the Sage. Otherwise, the Scribe coaches, then praises.

4. Students switch roles for the next problem.

ACTIVITY

13 GRAPH ME (STANDARD FORM)

▶ **Structure**
· Sage-N-Scribe

▶ **Materials**
· Transparency 8.2.8 and answer Transparency 8.2.8a
· 1 sheet of graph paper and pencil per pair

Graphic

Setup:
In pairs, Student A is the Sage; Student B is the Scribe. Students fold a sheet of paper in half and each writes his/her name on one half.

1. The Sage tells the Scribe how to graph the quadratic function in problem 1.

2. The Scribe graphs the parabola using the Sage's instructions.

3. If the Sage is correct, the Scribe praises the Sage. Otherwise, the Scribe coaches, then praises.

4. Students switch roles for the next problem.

ACTIVITY

14 REWRITE ME IN VERTEX FORM

▶ **Structure**
· Sage-N-Scribe

▶ **Materials**
· Transparency 8.2.9
· 1 sheet of paper and pencil per pair

Algebraic

Setup:
In pairs, Student A is the Sage; Student B is the Scribe. Students fold a sheet of paper in half and each writes his/her name on one half.

1. The Sage tells the Scribe how to rewrite the function in vertex form.

2. The Scribe records the Sage's instructions step-by-step in writing.

3. If the Sage is correct, the Scribe praises the Sage. Otherwise, the Scribe coaches, then praises.

4. Students switch roles for the next problem.

Cooperative Learning and Algebra 1: Becky Bride
Kagan Publishing • 1 (800) 933-2667 • www.KaganOnline.com

ACTIVITY

1 EXPLORING THE ROLE OF "A"

You will be investigating the effect the constant "a" has on the graph of a parabola whose equation is in the form of $y = a(x - h)^2 + k$. Use the vocabulary you have learned when appropriate.

Setup: Have students set their graphing calculator graph window to −12 to 12 for x and −12 to 12 for y.

1. The parent graph that will be the basis of all comparisons is $y = x^2$.
   a. What is the value of a in the parent graph? (What is the coefficient of x^2?)
   b. What is the value of h in the parent graph? (What is being subtracted from x?)
   c. What is the value of k in the parent graph? (What is being added to x^2?)

2. Graph $y = x^2$.

3. Graph $y = 2x^2$.
   a. Describe the effect 2 had on the graph when compared to the parent graph.
   b. Write the coordinates of the vertex.
   c. Write the equation of the axis of symmetry.
   d. Does the graph have a maximum or minimum point?
   e. Write the maximum or minimum value.
   f. Clear the graph $y = 2x^2$.

4. Graph $y = 5x^2$.
   a. Describe the effect 5 had on the graph when compared to the parent graph.
   b. Write the coordinates of the vertex.
   c. Write the equation of the axis of symmetry.
   d. Does the graph have a maximum or minimum point?
   e. Write the maximum or minimum value.
   f. Clear the graph $y = 5x^2$.

EXPLORING THE ROLE OF "A"

5. Graph $y = 9x^2$.

    a. Describe what effect 9 had on the graph when compared to the parent graph.

    b. Write the coordinates of the vertex.

    c. Write the equation of the axis of symmetry.

    d. Does the graph have a maximum or minimum point?

    e. Write the maximum or minimum value.

    f. Clear the graph $y = 9x^2$.

6. Describe the effect a has on the graph when a > 1.

7. Graph $y = 0.8x^2$.

    a. Describe the effect 0.8 had on the graph when compared to the parent graph.

    b. Write the coordinates of the vertex.

    c. Write the equation of the axis of symmetry.

    d. Does the graph have a maximum or minimum point?

    e. Write the maximum or minimum value.

    f. Clear the graph $y = 0.8x^2$.

8. Graph $y = 0.5x^2$.

    a. Describe the effect 0.5 had on the graph when compared to the parent graph.

    b. Write the coordinates of the vertex.

    c. Write the equation of the axis of symmetry.

    d. Does the graph have a maximum or minimum point?

    e. Write the maximum or minimum value.

    f. Clear the graph $y = 0.5x^2$.

Cooperative Learning and Algebra 1: Becky Bride
Kagan Publishing • 1 (800) 933-2667 • www.KaganOnline.com

ACTIVITY

1 EXPLORING THE ROLE OF "A"

9. Graph $y = 0.2x^2$.
   a. Describe the effect 0.2 had on the graph when compared to the parent graph.
   b. Write the coordinates of the vertex.
   c. Write the equation of the axis of symmetry.
   d. Does the graph have a maximum or minimum point?
   e. Write the maximum or minimum value.
   f. Clear the graph $y = 0.2x^2$.

10. Describe the effect a has on the graph when $0 < a < 1$.

11. Graph $y = -x^2$. Compare this to the parent graph.
    a. Describe the effect -1 had on the graph when compared to the parent graph.
    b. Write the coordinates of the vertex.
    c. Write the equation of the axis of symmetry.
    d. Does the graph have a maximum or minimum point?
    e. Write the maximum or minimum value.

12. Clear the graph $y = x^2$. The parent graph will now be $y = -x^2$.

13. Graph $y = -2x^2$.
    a. Describe the effect -2 had on the graph when compared to the parent graph.
    b. Write the coordinates of the vertex.
    c. Write the equation of the axis of symmetry.
    d. Does the graph have a maximum or minimum point?
    e. Write the maximum or minimum value.
    f. Clear the graph $y = -2x^2$.

EXPLORING THE ROLE OF "A"

14. Graph $y = -5x^2$.
   a. Describe the effect –5 had on the graph when compared to the parent graph.
   b. Write the coordinates of the vertex.
   c. Write the equation of the axis of symmetry.
   d. Does the graph have a maximum or minimum point?
   e. Write the maximum or minimum value.
   f. Clear the graph $y = -5x^2$.

15. Graph $y = -9x^2$.
   a. Describe the effect –9 had on the graph when compared to the parent graph.
   b. Write the coordinates of the vertex.
   c. Write the equation of the axis of symmetry.
   d. Does the graph have a maximum or minimum point?
   e. Write the maximum or minimum value.
   f. Clear the graph $y = -9x^2$.

16. Describe the effect a has on the graph when $a < -1$.

17. Graph $y = -0.8x^2$.
   a. Describe the effect –0.8 had on the graph when compared to the parent graph.
   b. Write the coordinates of the vertex.
   c. Write the equation of the axis of symmetry.
   d. Does the graph have a maximum or minimum point?
   e. Write the maximum or minimum value.
   f. Clear the graph $y = -0.8x^2$.

Cooperative Learning and Algebra 1: Becky Bride
Kagan Publishing • 1 (800) 933-2667 • www.KaganOnline.com

ACTIVITY

1 **EXPLORING THE ROLE OF "A"**

18. Graph $y = -0.5x^2$.
    a. Describe the effect -0.5 had on the graph when compared to the parent graph.
    b. Write the coordinates of the vertex.
    c. Write the equation of the axis of symmetry.
    d. Does the graph have a maximum or minimum point?
    e. Write the maximum or minimum value.
    f. Clear the graph $y = -0.5x^2$.

19. Graph $y = -0.2x^2$.
    a. Describe the effect -0.2 had on the graph when compared to the parent graph.
    b. Write the coordinates of the vertex.
    c. Write the equation of the axis of symmetry.
    d. Does the graph have a maximum or minimum point?
    e. Write the maximum or minimum value.
    f. Clear the graph $y = -0.2x^2$.

20. Describe the effect a has on the graph when $-1 < a < 0$.

21. Summarize the effect a has on the graph of a parabola, including a discussion about the direction it opens, vertex, axis of symmetry, maximum/minimum point, maximum/minimum value, vertical stretch, and vertical compression.

2, 4 & 6 WHAT DID I DO?

Structure: RallyRobin

Activity 2:
a) **How does the graph open?**
b) **Was the graph reflected over the x-axis?**
c) **Was there a vertical stretch or compression?**

Activity 4: Describe the horizontal shift.

Activity 6: Describe the vertical shift.

1. $y = 3(x - 4)^2 - 5$

2. $y = (-0.5)(x + 2)^2 + 4$

3. $y = (-0.25)x^2 + 1$

4. $2(x - 1)^2 = y$

5. $6 - 0.4(x + 3)^2 = y$

6. $8 + 2x^2 = y$

Answers on teachers notes (see pages 407, 408, and 409).

Cooperative Learning and Algebra 1: Becky Bride
Kagan Publishing • 1 (800) 933-2667 • www.KaganOnline.com

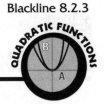

EXPLORING THE ROLE OF "H"

You will be investigating the effect the constant "h" has on the graph of a parabola whose equation is in the form of $y = a(x - h)^2 + k$. Use the vocabulary you have learned when appropriate.

Setup: Have students set their graphing calculator graph window to –12 to 12 for x and –12 to 12 for y.

1. The parent graph that will be the basis of all comparisons is $y = x^2$.
   a. What is the value of a in the parent graph? (What is the coefficient of x^2?)
   b. What is the value of h in the parent graph? (What is being subtracted from x?)
   c. What is the value of k in the parent graph? (What is being added to x^2?)

2. Graph $y = x^2$.

3. Graph $y = (x - 2)^2$. Compare this graph with the parent graph.
   a. What effect did subtracting 2 from x have on the graph when compared to the parent graph?
   b. Write the coordinates of the vertex.
   c. Write the equation of the axis of symmetry.
   d. Clear the graph $y = (x - 2)^2$.

4. Graph $y = (x - 7)^2$. Compare this graph with the parent graph.
   a. What effect did subtracting 7 from x have on the graph when compared to the parent graph?
   b. Write the coordinates of the vertex.
   c. Write the equation of the axis of symmetry.
   d. Clear the graph $y = (x - 7)^2$.

ACTIVITY

3 EXPLORING THE ROLE OF "H"

5. Graph $y = (x - 4)^2$. Compare this graph with the parent graph.
   a. What effect did subtracting 4 from x have on the graph when compared to the parent graph?
   b. Write the coordinates of the vertex.
   c. Write the equation of the axis of symmetry.
   d. Clear the graph $y = (x - 4)^2$.

6. Graph $y = (x + 5)^2$. Compare this graph with the parent graph.
   a. What effect did adding 5 to x have on the graph when compared to the parent graph?
   b. Write the coordinates of the vertex.
   c. Write the equation of the axis of symmetry.
   d. Clear the graph $y = (x + 5)^2$.

7. Graph $y = (x + 2)^2$. Compare this graph with the parent graph.
   a. What effect did adding 2 to x have on the graph when compared to the parent graph?
   b. Write the coordinates of the vertex.
   c. Write the equation of the axis of symmetry.
   d. Clear the graph $y = (x + 2)^2$.

8. Graph $y = (x + 6)^2$. Compare this graph with the parent graph.
   a. What effect did adding 6 to x have on the graph when compared to the parent graph?
   b. Write the coordinates of the vertex.
   c. Write the equation of the axis of symmetry.
   d. Clear the graph $y = (x + 6)^2$.

Cooperative Learning and Algebra 1: Becky Bride
Kagan Publishing • 1 (800) 933-2667 • www.KaganOnline.com

ACTIVITY
3

EXPLORING THE ROLE OF "H"

9. Summarize the effect h had on a graph when compared to the parent graph. Include in your summary:

a) The effect of adding a number to x (include a discussion about the coordinates of the vertex and the equation of the axis of symmetry).

b) The effect of subtracting a number from x (include a discussion about the coordinates of the vertex and the equation of the axis of symmetry).

ACTIVITY 5

EXPLORING THE ROLE OF "K"

You will be investigating the effect the constant "k" has on the graph of a parabola whose equation is in the form of $y = a(x - h)^2 + k$. Use the vocabulary you have learned when appropriate.

Setup: Have students set their graphing calculator graph window to -12 to 12 for x and -12 to 12 for y.

1. The parent graph that will be the basis of all comparisons is $y = x^2$.
   a. What is the value of a in the parent graph? (What is the coefficient of x^2?)
   b. What is the value of h in the parent graph? (What is being subtracted from x?)
   c. What is the value of k in the parent graph? (What is being added to x^2?)

2. Graph $y = x^2$.

3. Graph $y = x^2 + 3$. Compare this graph to the parent graph.
   a. What effect did adding 3 to x^2 have on the graph when compared to the parent graph?
   b. Write the coordinates of the vertex.
   c. Write the equation of the axis of symmetry.
   d. Write the maximum or minimum value.
   e. Clear the graph $y = x^2 + 3$.

4. Graph $y = x^2 + 6$. Compare this graph to the parent graph.
   a. What effect did adding 6 to x^2 have on the graph when compared to the parent graph?
   b. Write the coordinates of the vertex.
   c. Write the equation of the axis of symmetry.
   d. Write the maximum or minimum value.
   e. Clear the graph $y = x^2 + 6$.

Cooperative Learning and Algebra 1: Becky Bride
Kagan Publishing • 1 (800) 933-2667 • www.KaganOnline.com

EXPLORING THE ROLE OF "K"

5. Graph $y = x^2 + 2$. Compare this graph to the parent graph.
   a. What effect did adding 2 to x^2 have on the graph when compared to the parent graph?
   b. Write the coordinates of the vertex.
   c. Write the equation of the axis of symmetry.
   d. Write the maximum or minimum value.
   e. Clear the graph $y = x^2 + 2$.

6. Graph $y = x^2 - 5$. Compare this graph to the parent graph.
   a. What effect did subtracting 5 from x^2 have on the graph when compared to the parent graph?
   b. Write the coordinates of the vertex.
   c. Write the equation of the axis of symmetry.
   d. Write the maximum or minimum value.
   e. Clear the graph $y = x^2 - 5$.

7. Graph $y = x^2 - 1$. Compare this graph to the parent graph.
   a. What effect did subtracting 1 from x^2 have on the graph when compared to the parent graph?
   b. Write the coordinates of the vertex.
   c. Write the equation of the axis of symmetry.
   d. Write the maximum or minimum value.
   e. Clear the graph $y = x^2 - 1$.

8. Graph $y = x^2 - 4$. Compare this graph to the parent graph.
   a. What effect did subtracting 4 from x^2 have on the graph when compared to the parent graph?
   b. Write the coordinates of the vertex.
   c. Write the equation of the axis of symmetry.
   d. Write the maximum or minimum value.
   e. Clear the graph $y = x^2 - 4$.

ACTIVITY

5 **EXPLORING THE ROLE OF "K"**

9. Summarize the effect k has on a graph when compared to the parent graph. Include in your summary:
    a) The effect of adding a number to x^2 (include a discussion about the coordinates of the vertex, the maximum or minimum value, and the equation of the axis of symmetry).
    b) The effect of subtracting a number from x^2 (include a discussion about the coordinates of the vertex, the maximum or minimum value, and the equation of the axis of symmetry).

The next part of the investigation has you explore the graphs of quadratic functions that contain both h and k.

10. Clear all graphs.

11. Graph $y = (x - 3)^2 + 1$.
    a. Describe the horizontal shift of the graph when compared to $y = x^2$.
    b. Describe the vertical shift of the graph when compared to $y = x^2$.
    c. What are the coordinates of the vertex?
    d. What is the equation of the axis of symmetry?
    e. What is the minimum or maximum value?

12. Graph $y = (x + 2)^2 - 3$.
    a. Describe the horizontal shift of the graph when compared to $y = x^2$.
    b. Describe the vertical shift of the graph when compared to $y = x^2$.
    c. What are the coordinates of the vertex?
    d. What is the equation of the axis of symmetry?
    e. What is the minimum or maximum value?

13. Graph $y = (x - 1)^2 + 5$.
    a. Describe the horizontal shift of the graph when compared to $y = x^2$.
    b. Describe the vertical shift of the graph when compared to $y = x^2$.
    c. What are the coordinates of the vertex?

Cooperative Learning and Algebra 1: Becky Bride
Kagan Publishing • 1 (800) 933-2667 • www.KaganOnline.com

    d. What is the equation of the axis of symmetry?

    e. What is the minimum or maximum value?

14. Graph $y = (x + 6)^2 - 2$.

    a. Describe the horizontal shift of the graph when compared to $y = x^2$.

    b. Describe the vertical shift of the graph when compared to $y = x^2$.

    c. What are the coordinates of the vertex?

    d. What is the equation of the axis of symmetry?

    e. What is the minimum or maximum value?

15. DO NOT GRAPH $y = (x + 13)^2 - 16$. Based on your observations in problems 11-14,

    a. Describe the horizontal shift of the graph when compared to $y = x^2$.

    b. Describe the vertical shift of the graph when compared to $y = x^2$.

    c. What are the coordinates of the vertex?

    d. What is the equation of the axis of symmetry?

    e. What is the minimum or maximum value?

16. DO NOT GRAPH $y = (x - 17)^2 + 20$. Based on your observations in problems 11-14,

    a. Describe the horizontal shift of the graph when compared to $y = x^2$.

    b. Describe the vertical shift of the graph when compared to $y = x^2$.

    c. What are the coordinates of the vertex?

    d. What is the equation of the axis of symmetry?

    e. What is the minimum or maximum value?

7 & 9 GRAPH ME (VERTEX FORM)
REWRITE ME IN STANDARD FORM

Structure: Sage-N-Scribe

Activity 7: Graph each function.
Activity 9: Rewrite each function in standard form.

1. $y = (x - 1)^2 + 3$

2. $y = -(x + 2)^2 - 4$

3. $y = -(x + 3)^2 - 1$

4. $y = (x - 2)^2 + 5$

Answers for Activity 9:
1. $x^2 - 2x + 4$
2. $-x^2 - 4x - 8$
3. $-x^2 - 6x - 10$
4. $x^2 - 4x + 9$

Cooperative Learning and Algebra 1: Becky Bride
Kagan Publishing • 1 (800) 933-2667 • www.KaganOnline.com

QUADRATIC FUNCTIONS

Answers for Activity 7

1.

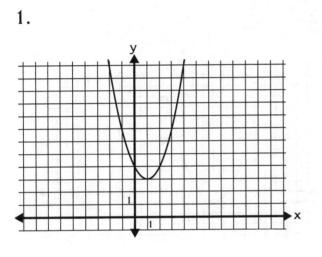

2.

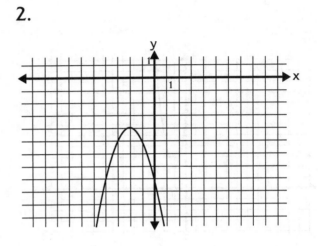

3.

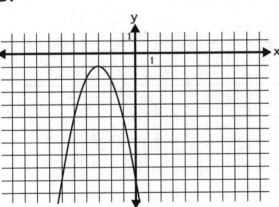

4.

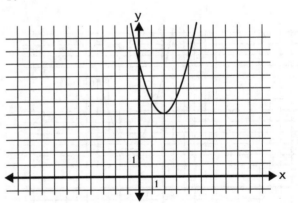

WRITE MY EQUATION

Structure: Sage-N-Scribe

Write the equation of each quadratic function below.

1.

2.

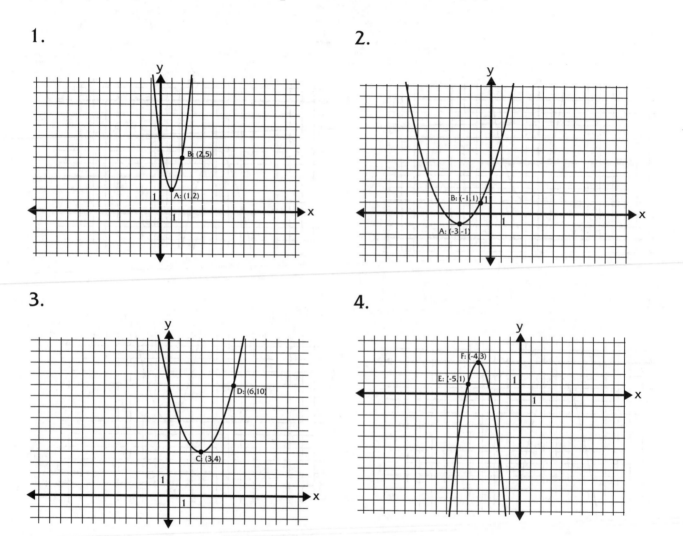

3.

4.

Answers are on the teacher notes (see page 410).

Cooperative Learning and Algebra 1: Becky Bride
Kagan Publishing • 1 (800) 933-2667 • www.KaganOnline.com

10, 12 & 13 — WHAT ARE THE VALUES OF A, B, AND C?

Structure: RallyRobin & Sage-N-Scribe

Activity 10 directions: State the value of a, b, and c.
Activity 12 directions: Find the coordinates of the vertex.
Activity 13 directions: Graph each function.

1. $y = x^2 - 2x - 7$

2. $y = x^2 + 4x - 2$

3. $y = -3x^2 + 6x - 4$

4. $y = 2x^2 + 8x - 3$

Answers:

Activity 10:
1. 1; −2; −7
2. 1; 4; −2
3. −3; 6; −4
4. 2; 8; 3

Activity 13:
See page 433.

Activity 12:
1. (1, −8)
2. (−2, −6)
3. (1, −1)
4. (−2, −5)

EXPLORING THE VERTEX (STANDARD FORM)

1. What is the vertex of the parabola whose equation is $y = 3(x - 1)^2 + 4$?

2. For the parabola whose equation is $y = 3x^2 - 6x + 7$, find the value of $\dfrac{-b}{2a}$.

3. Take your answer from question 2, and substitute it for x into the function $y = 3x^2 - 6x + 7$. Find the corresponding y-value and write as an ordered pair.

4. Compare the ordered pair in question 3 with the vertex in question 1. What is similar?

5. Rewrite the equation in question 1 in standard form.

6. Compare your answer in question 5 to the equation in question 2. Do the two equations represent the same parabola? If you are not sure, graph both on your graphing calculator.

7. What is the vertex of the parabola whose equation is $y = 0.5 (x - 2)^2 - 3$?

Cooperative Learning and Algebra 1: Becky Bride
Kagan Publishing • 1 (800) 933-2667 • www.KaganOnline.com

EXPLORING THE VERTEX
(STANDARD FORM)

8. For the parabola whose equation is $y = 0.5x^2 - 2x - 1$, find the value of $\dfrac{-b}{2a}$.

9. Take your answer from question 8, and substitute it for x into the function $y = 0.5x^2 - 2x - 1$. Find the corresponding y-value and write as an ordered pair.

10. Compare the ordered pair in question 9 with the vertex in question 7. What is similar?

11. Rewrite the equation in question 7 in standard form.

12. Compare your answer in question 11 to the equation in question 8. Do the two equations represent the same parabola? If you are not sure, graph both on your graphing calculator.

13. What is the vertex of the parabola whose equation is $y = (x + 5)^2 + 2$?

14. For the parabola whose equation is $y = x^2 + 10x + 27$, find the value of $\dfrac{-b}{2a}$.

EXPLORING THE VERTEX (STANDARD FORM)

15. Take your answer from question 14, and substitute it for x into the function $y = x^2 + 10x + 27$. Find the corresponding y-value and write as an ordered pair.

16. Compare the ordered pair in question 15 with the vertex in question 13. What is similar?

17. Rewrite the equation in question 13 in standard form.

18. Compare your answer in question 17 to the equation in question 14. Do the two equations represent the same parabola? If you are not sure, graph both on your graphing calculator.

19. Explain how to find the vertex of a parabola whose equation is in standard form, based on the investigation just completed.

20. Apply your new knowledge by finding the vertex of the parabola whose equation is $y = 2x^2 + 12x + 17$.

Cooperative Learning and Algebra 1: Becky Bride
Kagan Publishing • 1 (800) 933-2667 • www.KaganOnline.com

ACTIVITY

13 GRAPH ME (STANDARD FORM)

QUADRATIC FUNCTIONS

Structure: Sage-N-Scribe

Answers for Activity 13.

1.

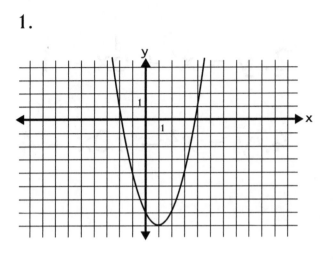

2.

3.

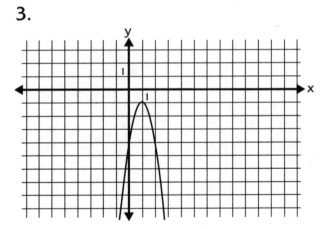

4.

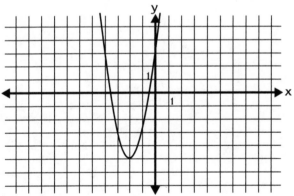

Cooperative Learning and Algebra 1: Becky Bride
Kagan Publishing • 1 (800) 933-2667 • www.KaganOnline.com

14 REWRITE ME IN VERTEX FORM

Structure: Sage-N-Scribe

Rewrite each in vertex form.

1. $y = x^2 - 4x + 5$

2. $y = x^2 + 6x - 1$

3. $y = x^2 + 8x + 23$

4. $y = x^2 - 10x + 4$

Answers:
1. $y = (x - 2)^2 + 1$
2. $y = (x + 3)^2 - 10$
3. $y = (x + 4)^2 + 7$
4. $y = (x - 5)^2 - 21$

Cooperative Learning and Algebra 1: Becky Bride
Kagan Publishing • 1 (800) 933-2667 • www.KaganOnline.com

LESSON 3
SOLVING QUADRATIC FUNCTIONS

This lesson focuses on finding solutions to quadratic functions. It begins with an algebraic-geometric connection. Students solved quadratic functions using factoring in the last chapter. Activity 2 recommends that the students use factoring and then the quadratic formula for two reasons: 1) it reinforces factoring as a method of finding solutions and 2) it reinforces that the quadratic formula finds the same solutions as the factoring method. Activity 3 has irrational and complex answers. Students will not actually give the complex answers, but they need to know that nonreal answers can occur. This leads into the discriminant and the lesson ends with applications and a synthesis activity.

ACTIVITY 1
EXPLORING THE CONNECTION: QUADRATIC FORMULA AND X-INTERCEPTS

Exploratory

Setup:
Have students make their graphing calculator window be –10 to 10 for x-values and –10 to 10 for y-values.

Solo

1. Have each student complete the investigation individually.

Pair Consensus

2. For each problem on the investigation, each student shares with his/her partner, using RallyRobin, his/her response. They discuss the problems that they disagree on, trying to come to consensus on the correct response. For the problems they can't reach consensus on, they mark these so they can focus on them during the team phase. Encourage the students to add to their responses if their partner verbalized an understanding they did not see.

Team Consensus

3. Each pair shares their responses, using RallyRobin, with the other pair in their team, augmenting their responses if necessary. When the teams are through sharing, each student should have a detailed, complete summary.

Tip: You may want to review that square roots of negative numbers cannot be found. This is critical to the last 14 problems on the investigation.

▶ **Structure**
• Solo-Pair Consensus-Team Consensus

▶ **Materials**
• 1 Blackline 8.3.1 per student
• 1 sheet of paper, pen/pencil, and graphing calculator per student

2 FIND MY SOLUTION(S)

▶ **Structure**
· Sage-N-Scribe

▶ **Materials**
· Transparency 8.3.2
· 1 sheet of paper and pencil per pair

Algebraic

Setup:
In pairs, Student A is the Sage; Student B is the Scribe. Students fold a sheet of paper in half and each writes his/her name on one half.

1. The Sage tells the Scribe how to solve the quadratic function in problem 1.

2. The Scribe records the Sage's instructions step-by-step in writing.

3. If the Sage is correct, the Scribe praises the Sage. Otherwise, the Scribe coaches, then praises.

4. Students switch roles for the next problem.

Tip: Have students do this activity first using the factoring method. Then have the students redo this activity using the quadratic formula.

3 FIND MY SOLUTION(S) AND SIMPLIFY

▶ **Structure**
· Sage-N-Scribe

▶ **Materials**
· Transparency 8.3.3
· 1 sheet of paper and pencil per pair

Algebraic

Setup:
In pairs, Student A is the Sage; Student B is the Scribe. Students fold a sheet of paper in half and each writes his/her name on one half.

1. The Sage tells the Scribe how to solve the quadratic function, using the quadratic formula, in problem 1.

2. The Scribe records the Sage's instructions step-by-step in writing.

3. If the Sage is correct, the Scribe praises the Sage. Otherwise, the Scribe coaches, then praises.

4. Students switch roles for the next problem.

Cooperative Learning and Algebra 1: Becky Bride
Kagan Publishing · 1 (800) 933-2667 · www.KaganOnline.com

4 EXPLORING THE DISCRIMINANT

Exploratory

Solo

1. Have each student complete the investigation individually.

Pair Consensus

2. For each problem on the investigation, each student shares with his/her partner, using RallyRobin, his/her response. They discuss the problems that they disagree on, trying to come to consensus on the correct response. For the problems they can't reach consensus on, they mark these so they can focus on them during the team phase. Encourage the students to add to their responses if their partner verbalized an understanding they did not see.

Team Consensus

3. Each pair shares their responses, using RallyRobin, with the other pair in their team, augmenting their responses if necessary. When the teams are through sharing, each student should have a detailed, complete summary.

▶ **Structure**
· Solo-Pair Consensus-Team Consensus

▶ **Materials**
· 1 Blackline 8.3.4 per student
  1 sheet of paper and pen/pencil per student

5 HOW MANY SOLUTIONS DO I HAVE?

Algebraic

Setup:
In pairs, Student A is the Sage; Student B is the Scribe. Students fold a sheet of paper in half and each writes his/her name on one half.

1. The Sage tells the Scribe how to use the discriminant to determine the number of solutions for problem 1.

2. The Scribe records the Sage's instructions step-by-step in writing.

3. If the Sage is correct, the Scribe praises the Sage. Otherwise, the Scribe coaches, then praises.

4. Students switch roles for the next problem.

▶ **Structure**
· Sage-N-Scribe

▶ **Materials**
· Transparency 8.3.5
· 1 sheet of paper and pencil per pair

ACTIVITY

6

APPLICATIONS

▶ **Structure**
• Sage-N-Scribe

▶ **Materials**
• Transparency 8.3.6
• 1 sheet of paper and pencil per pair

Graphical

Setup:
In pairs, Student A is the Sage; Student B is the Scribe. Students fold a sheet of paper in half and each writes his/her name on one half.

1. The Sage tells the Scribe how to solve problem 1.

2. The Scribe records the Sage's instructions step-by-step in writing.

3. If the Sage is correct, the Scribe praises the Sage. Otherwise, the Scribe coaches, then praises.

4. Students switch roles for the next problem.

ACTIVITY

7

WHAT DID WE LEARN?

▶ **Structure**
• RoundTable Consensus

▶ **Materials**
• 1 large sheet of paper per team
• different color pen or pencil for each student on the team

Synthesis via Graphic Organizer

1. Each teammate signs his/her name in the upper right corner of the team paper with the color pen/pencil he/she is using.

2. One teammate writes "Quadratic Functions" in the center of the team paper in a rectangle.

3. Teammate 1 shares with the team one core concept he/she learned in the unit.

4. The student checks for consensus.

5. The teammates show agreement or lack of agreement with thumbs up or down.

6. If there is agreement, the students celebrate and the teammate records the core concept on the graphic organizer connecting it with a line to the main idea, "Quadratic Functions." If not, teammates discuss the response until there is agreement and then they celebrate.

7. Play continues with the next student's core concept until all core concepts are exhausted.

8. Repeat steps 3-7, with teammates adding details to each core concept and making bridges between related ideas.

Cooperative Learning and Algebra 1: Becky Bride
Kagan Publishing • 1 (800) 933-2667 • www.KaganOnline.com

ACTIVITY 1 EXPLORING THE CONNECTION: QUADRATIC FORMULA AND X-INTERCEPTS

1. Look at the graph to the right. What are the x-intercepts of the graph?

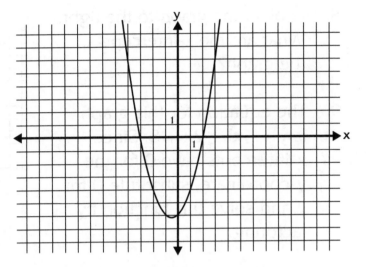

2. The equation for this graph is $y = x^2 + x - 6$. Since x-intercepts have 0 for the y-coordinate, substitute 0 for y and solve this function by factoring.

3. Compare the answers in problem 2 with the x-intercepts in problem 1. What do you notice?

4. Write the values of a, b, and c for the quadratic function in problem 2.

5. Substitute these values into the expression $x = \dfrac{-b + \sqrt{b^2 - 4ac}}{2a}$. Show all work below.

6. Substitute these values into the expression $x = \dfrac{-b - \sqrt{b^2 - 4ac}}{2a}$. Show all work below.

7. Compare your answers to problems 5 and 6 with problems 1 and 2. What do you notice?

8. Look at the graph to the right. What are the x-intercepts of the graph?

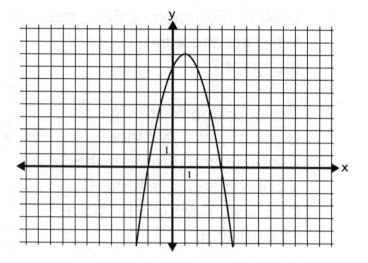

9. The equation for this graph is $y = -x^2 + 2x + 8$. Since x-intercepts have 0 for the y-coordinate, substitute 0 for y and solve this function by factoring.

10. Compare the answers in problem 9 with the x-intercepts in problem 8. What do you notice?

11. Write the values of a, b, and c for the quadratic function in problem 9.

12. Substitute these values into the expression $x = \dfrac{-b + \sqrt{b^2 - 4ac}}{2a}$. Show all work below.

13. Substitute these values into the expression $x = \dfrac{-b - \sqrt{b^2 - 4ac}}{2a}$. Show all work below.

14. Compare your answers to problems 12 and 13 with problems 8 and 9. What do you notice?

Cooperative Learning and Algebra 1: Becky Bride
Kagan Publishing • 1 (800) 933-2667 • www.KaganOnline.com

EXPLORING THE CONNECTION: QUADRATIC FORMULA AND X-INTERCEPTS

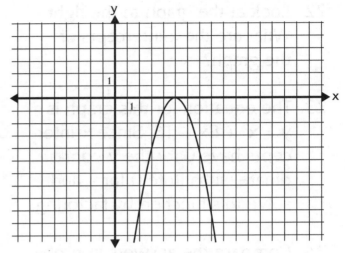

15. Look at the graph to the right. What are the x-intercepts of the graph?

16. The equation for this graph is $y = -x^2 + 10x - 25$. Since x-intercepts have 0 for the y-coordinate, substitute 0 for y and solve this function by factoring.

17. Compare the answers in problem 16 with the x-intercepts in problem 15. What do you notice?

18. Write the values of a, b, and c for the quadratic function in problem 16.

19. Substitute these values into the expression $x = \dfrac{-b + \sqrt{b^2 - 4ac}}{2a}$. Show all work below.

20. Substitute these values into the expression $x = \dfrac{-b - \sqrt{b^2 - 4ac}}{2a}$. Show all work below.

21. Compare your answers to problems 19 and 20 with problems 15 and 16. What do you notice?

ACTIVITY 1 EXPLORING THE CONNECTION: QUADRATIC FORMULA AND X-INTERCEPTS

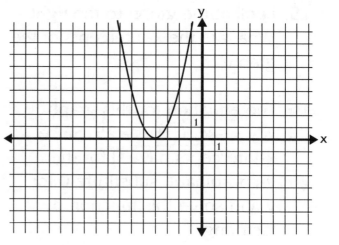

22. Look at the graph to the right. What are the x-intercepts of the graph?

23. The equation for this graph is $y = x^2 + 8x + 16$. Since x-intercepts have 0 for the y-coordinate, substitute 0 for y and solve this function by factoring.

24. Compare the answers in problem 23 with the x-intercepts in problem 22. What do you notice?

25. Write the values of a, b, and c for the quadratic function in problem 23.

26. Substitute these values into the expression $x = \dfrac{-b + \sqrt{b^2 - 4ac}}{2a}$. Show all work below.

27. Substitute these values into the expression $x = \dfrac{-b - \sqrt{b^2 - 4ac}}{2a}$. Show all work below.

28. Compare your answers to problems 26 and 27 with problems 22 and 23. What do you notice?

29. Reread your answers for problems 7, 14, 21, and 28. Describe two ways that x-intercepts can be found algebraically.

Cooperative Learning and Algebra 1: Becky Bride
Kagan Publishing • 1 (800) 933-2667 • www.KaganOnline.com

EXPLORING THE CONNECTION: QUADRATIC FORMULA AND X-INTERCEPTS

30. Look at the graph to the right. What are the x-intercepts of the graph?

31. The equation for this graph is $y = x^2 + x + 3$. Since x-intercepts have 0 for the y-coordinate, substitute 0 for y and solve this function by factoring. Can it be factored?

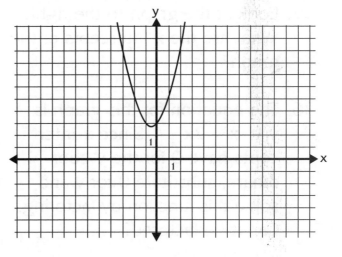

32. Substitute these values into the expression $x = \dfrac{-b + \sqrt{b^2 - 4ac}}{2a}$. Show all work below.

33. Substitute these values into the expression $x = \dfrac{-b - \sqrt{b^2 - 4ac}}{2a}$. Show all work below.

34. Compare your answers to problems 32 and 33 with the x-intercepts in problem 30. What do you notice?

EXPLORING THE CONNECTION:
QUADRATIC FORMULA AND X-INTERCEPTS

35. Look at the graph to the right. What are the x-intercepts of the graph?

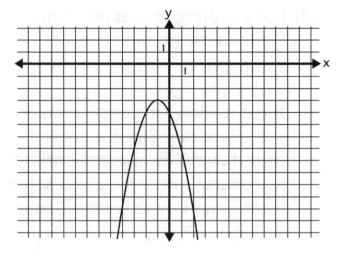

36. The equation for this graph is $y = -x^2 - 2x - 4$. Since x-intercepts have 0 for the y-coordinate, substitute 0 for y and solve this function by factoring. Can it be factored?

37. Substitute these values into the expression $x = \dfrac{-b + \sqrt{b^2 - 4ac}}{2a}$. Show all work below.

38. Substitute these values into the expression $x = \dfrac{-b - \sqrt{b^2 - 4ac}}{2a}$. Show all work below.

39. Compare your answers to problems 37 and 38 with the x-intercepts in problem 35. What do you notice?

444

40. The equation $x = \dfrac{-b \pm \sqrt{b^2 - 4ac}}{2a}$ is called the "quadratic formula." It is an algebraic way of finding x-intercepts and can be used whether the quadratic is factorable or not. The last two graphs did not have x-intercepts. Re-examine problems 32, 33, 37, and 38. Compare these with problems 5, 6, 12 ,13, 19, 20, 26, and 27 and explain how you can look at the quadratic formula after values have been substituted and know that there are no x-intercepts.

41. The x-intercepts are also called solutions of the quadratic function. Looking at all of the graphs in this investigation, what are the possible number of solutions of a quadratic function?

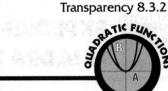

ACTIVITY 2
FIND MY SOLUTION(S)

Structure: Sage-N-Scribe

Find the solutions.

1. $y = 2x^2 + 7x + 3$

2. $y = 5x^2 + 2x - 3$

3. $36x^2 + 12x + 1 = y$

4. $4x^2 - 12x + 9 = y$

5. $4x^2 - 1 = y$

6. $3x^2 + 4x = y$

Answers:

1. $\frac{-1}{2}$ and -3

2. $\frac{3}{5}$ and -1

3. $\frac{-1}{6}$

4. $\frac{3}{2}$

5. $\frac{1}{2}$ and $\frac{-1}{2}$

6. $\frac{-4}{3}$ and 0

Cooperative Learning and Algebra 1: Becky Bride
Kagan Publishing • 1 (800) 933-2667 • www.KaganOnline.com

FIND MY SOLUTION(S)
AND SIMPLIFY

Structure: Sage-N-Scribe

Solve each quadratic function using the quadratic formula.

1. $y = 4x^2 - 6x + 1$ 2. $y = 2x^2 - 4x - 3$

3. $3x^2 - x + 4 = y$ 4. $3x^2 + 4x - 3 = y$

Answers:

1. $\dfrac{3+\sqrt{5}}{4}$ and $\dfrac{3-\sqrt{5}}{4}$ or 1.31 and 0.19

2. $\dfrac{2+\sqrt{10}}{2}$ and $\dfrac{2-\sqrt{10}}{2}$ or 2.58 and -0.58

3. no real solution

4. $\dfrac{-2+\sqrt{13}}{3}$ and $\dfrac{-2-\sqrt{13}}{3}$ or 0.54 and -1.87

ACTIVITY

4 EXPLORING THE DISCRIMINANT

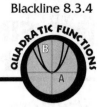

Graph 1

Graph 2

Graph 3

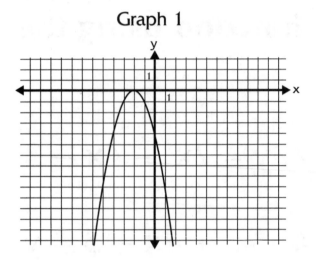

Graph 4

Graph 5

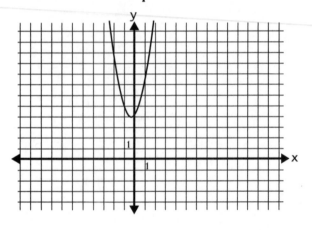

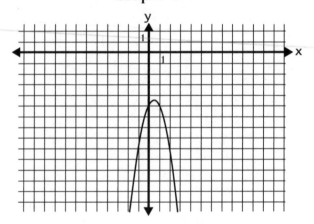

Graph 6

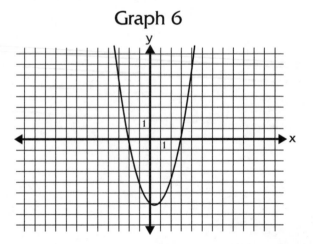

Cooperative Learning and Algebra 1: Becky Bride
Kagan Publishing • 1 (800) 933-2667 • www.KaganOnline.com

4 EXPLORING THE DISCRIMINANT

1. Solutions to a quadratic function are the x-intercepts of its graph. Write the number of solutions for:
   a) Graph 1   b) Graph 2   c) Graph 3

   d) Graph 4   e) Graph 5   f) Graph 6

2. The quadratic function for graph 1 is $y = -x^2 - 4x - 4$. Substitute the values for a, b, and c into the expression $b^2 - 4ac$. Is the answer a positive number, negative number, or zero? Refer back to problem 1. How many solutions did the function have?

3. The quadratic function for graph 2 is $y = -4x^2 + 4x - 1$. Substitute the values for a, b, and c into the expression $b^2 - 4ac$. Is the answer a positive number, negative number, or zero? Refer back to problem 1. How many solutions did the function have?

4. The quadratic function for graph 3 is $y = 2x^2 + x + 4$. Substitute the values for a, b, and c into the expression $b^2 - 4ac$. Is the answer a positive number, negative number, or zero? Refer back to problem 1. How many solutions did the function have?

5. The quadratic function for graph 4 is $y = -x^2 + x - 5$. Substitute the values for a, b, and c into the expression $b^2 - 4ac$. Is the answer a positive number, negative number, or zero? Refer back to problem 1. How many solutions did the function have?

4 EXPLORING THE DISCRIMINANT

6. The quadratic function for graph 5 is $y = -2x^2 - 8x - 3$. Substitute the values for a, b, and c into the expression $b^2 - 4ac$. Is the answer a positive number, negative number, or zero? Refer back to problem 5. How many solutions did the function have?

7. The quadratic function for graph 6 is $y = x^2 - x - 6$. Substitute the values for a, b, and c into the expression $b^2 - 4ac$. Is the answer a positive number, negative number, or zero? Refer back to problem 6. How many solutions did the function have?

8. The expression $b^2 - 4ac$ is called the "discriminant." It is part of the quadratic formula. Where in the quadratic formula is the discriminant located?

9. The discriminant is a quick and simple algebraic method to determine how many solutions a quadratic function has. Re-examine the answers to problems 2-7. Then summarize how to use the discriminant to determine the number of solutions of a quadratic function.

Cooperative Learning and Algebra 1: Becky Bride
Kagan Publishing • 1 (800) 933-2667 • www.KaganOnline.com

HOW MANY SOLUTIONS DO I HAVE?

Structure: Sage-N-Scribe

Determine the type and number of solutions each function has using the discriminant.

1. $y = 4x^2 - x + 3$

2. $y = 4x^2 - 5x - 1$

3. $2x^2 - 3x - 9 = y$

4. $25x^2 + 10x + 1 = y$

5. $y = x^2 - 4x - 6$

6. $5x^2 + 2x + 3 = y$

Answers:
1. no real solutions
2. 2 irrational solutions
3. 2 rational solutions
4. 1 real solution
5. 2 irrational solutions
6. no real solutions

6 APPLICATIONS

Structure: Sage-N-Scribe

Solve each problem.

For problems 1-2, use the following situation.
If a model rocket is launched from a 4-foot platform with an initial velocity of 100 feet per second, then its height can be modeled by $y = -16x^2 + 100x + 4$, where x represents the time in seconds since the launch and y represents the height in feet.

1. After how many seconds did the rocket hit the ground? Round to the hundredths place.

2. What was the greatest height the rocket reached? Round to the hundredths place.

For problems 3-4, use the following situation.
The number of seeds dispersed by a plot of dandelions each day in June can be modeled by $y = 20x - x^2$, where x is the number of days in June and y represents the number of seeds dispersed.

3. What was the greatest number of seeds dispersed during the month of June? Round to the hundredths place.

4. On what day of the month were seeds no longer being dispersed?

Cooperative Learning and Algebra 1: Becky Bride
Kagan Publishing • 1 (800) 933-2667 • www.KaganOnline.com

6 APPLICATIONS

Structure: Sage-N-Scribe

For problems 5-6, use the following situation.

The physics classes built catapults to launch pumpkins. If a pumpkin is launched from ground level with an initial velocity of 30 feet per second, then its height can be modeled by $y = 30x - 16x^2$, where x is the number of seconds after launch and y is the height of the pumpkin in feet.

5. What is the greatest height the pumpkin reaches? Round to the hundredths place.

6. How high above the ground is the pumpkin after 1 second has passed?

Answers:
1. 6.28 seconds
2. 160.25 feet
3. 100 seeds
4. June 20
5. 14.06 feet
6. 14 feet